Gender Revealed
Becoming Male or Female

Gender Revealed
Becoming Male or Female

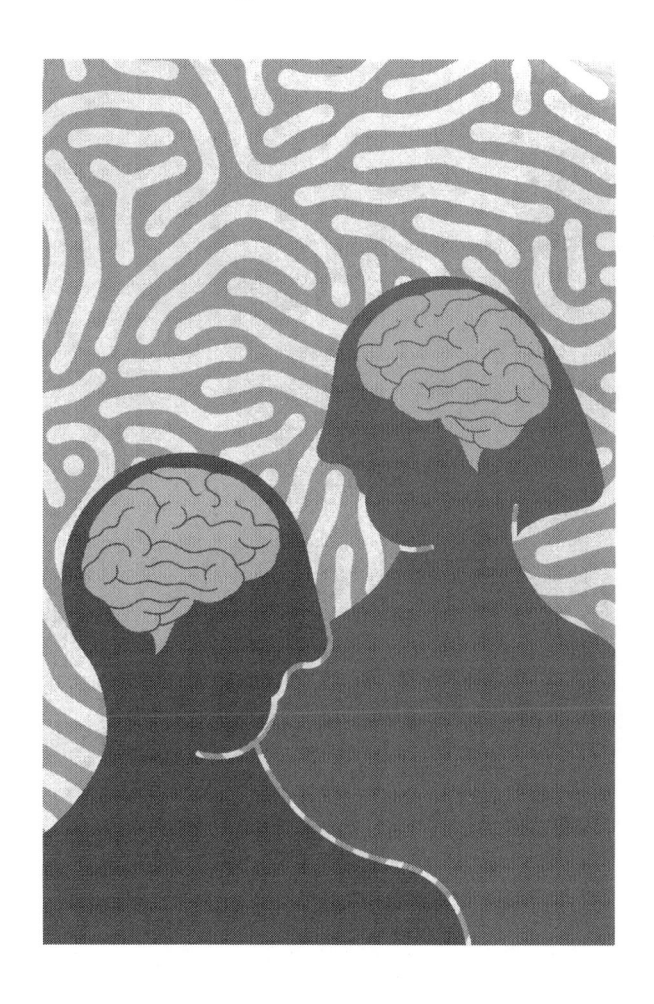

Daniel Y. Parkinson, MD

Copyright © 2024 Daniel Y. Parkinson

All rights reserved

No part of this book may be reproduced, or stored in a retrieval system, or transmitted in any form or by any means, electronic, mechanical, photocopying, recording, or otherwise, without express written permission of the publisher or author.

ISBN: 979-8-9917404-1-8

Cover design and illustrations by Diego Villalobos Ulate

Printed in the United States of America

Table of Contents

Chapter 1—Intro and Definitions   9

Chapter 2—Chromosomes part 1   22

Chapter 3—Testosterone   28

Chapter 4—Apps   58

Chapter 5—Orientation and Identity   99

Chapter 6—Chromosomes part 2   132

Chapter 7—Epilogue   142

Chapter 1—Intro and Definitions

Intro

This book is about sex, gender, and the biological processes that make most humans either a male or a female.

Most people, including most physicians, don't really understand sex and gender much beyond X and Y chromosomes, genitalia, gonads, and sex hormones. Beyond that it seems like an enormous mystery. In reality, there is enough knowledge available to really understand the main mechanisms of how it all comes together. However, that knowledge is spread out between various disciplines from embryology, genetics, developmental biology, neurology, neuroanatomy, endocrinology, neuropsychology and many more. To get the big picture, we need a synthesis of the knowledge found in these various disciplines. I have not yet found that synthesis, so I am putting one forward. It is my goal to pull together the knowledge from these disciplines into a coherent narrative.

There is currently a lot of discussion going on about sex and gender, and a lot of this is spilling over into politics. Unfortunately, much of the discussion is misinformed. The discussion would have a better outcome if it was based on knowledge and understanding. Public policies, personal opinions, and political debates will improve if all sides are arguing from an informed position.

I will be presenting some complex concepts in the following chapters. However, understanding these ideas will not require a substantial background in science. Hopefully, this narrative will be understandable to most everybody. I am focusing on the broader patterns, and I will avoid small details or highly specific scientific terminology, so that I can focus on the big picture.

I tried to keep this book very concise. It is written to inform, but it is also an incredibly fascinating subject, and I want to make this information as accessible as possible. Although some of the concepts in the following chapters are complex, they are all understandable, even without a science background, and some of the nuances are important—not to mention—incredibly interesting.

Defining the nouns "male" and "female"

In order for the conversation to make any sense, we need to agree on some terms.

When we use words like "male" or "female," usually we can all agree on the meanings without needing to define them. However, when you start looking at the biology, you realize that these words can be defined in many different ways, and occasionally, those different definitions can be incomplete, inconsistent, or even contradictory.

Chapter 1—Intro and Definitions

Let's review some different systems we can use to define male/female and what criteria these systems use.

Definition/system #1—potential role in procreation

In English, for hundreds of years, we have been using terms like female, male, man, woman, girl, boy, etc., and we have pretty much agreed on what we mean. The common wisdom has been that "female" refers to anybody who appears to have the body parts necessary to get pregnant and deliver a child at some point in their life, while "male" refers to anybody who appears to have the body parts necessary to impregnate a female at some point in their life.

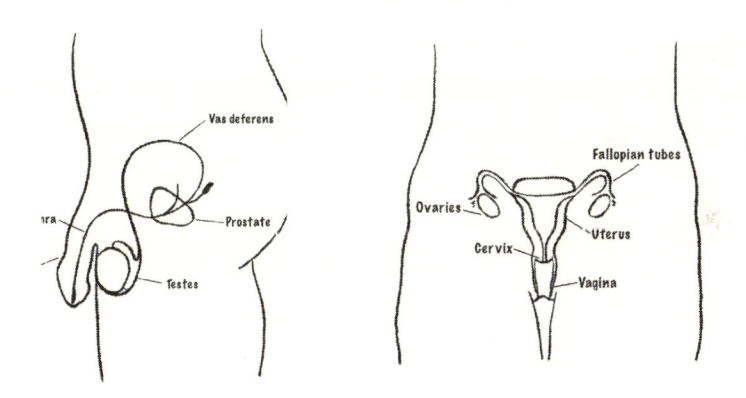

Definition/system #2—presence or absence of penis

A slightly different classification has been to base sex on the presence or absence of a penis. When each baby is born, they are assumed to be male or female based on whether they

11

appear to have a penis vs. a vagina. (Confusion has ensued in those rare cases where the newborn had both or neither.)

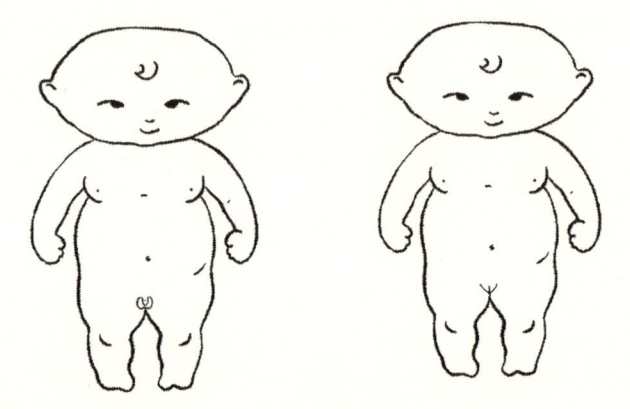

For eons, this has been the default system for assigning gender to infants. This system works pretty well most of the time, and we still use it. Nowadays, most infants in the USA are assigned a gender even before they are born based on whether or not a penis is detected on an ultrasound.

Definition/system #3—appearance and presentation

In modern society, people cover up and rarely allow others to observe their genitals. On a day-to-day basis, we use other factors to try to distinguish males from females. We consider physical traits that we can observe, even though most of these traits don't directly impact procreation. However, these traits are considered to be part of a package of femaleness or maleness. They include things like hair patterns, pitch of voice, body shape and size—traits that mostly emerge with puberty.

Chapter 1—Intro and Definitions

Some other important clues have nothing to do with physical features. Most of the biggest signals of maleness or femaleness are completely arbitrary and culture-specific. I am referring to how a person presents themselves (or how they are presented by their parents in the case of young children). The way a person dresses, styles their hair, and decorates their face/body signals a person's sex, even when no physical characteristics are otherwise visible.

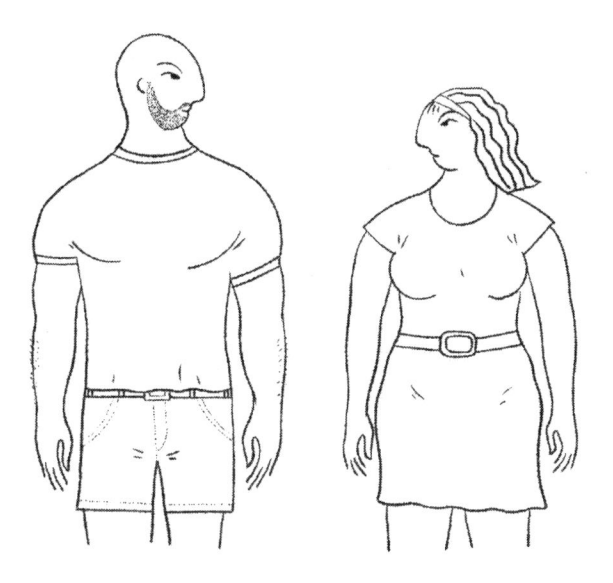

So, this system defines gender based on differences we can detect in routine, everyday encounters. Even though it seems less definitive, appearance/presentation is actually the system we use the most.

Every time we encounter a person, we gain an impression of their sex based on what we can easily see (and hear). We make assumptions based on the physical characteristics that we can observe, as well as on their presentation (clothing, hair, etc.) and then we subsequently expect them to behave in certain ways. We expect people to talk and act in gender-specific ways, based on how they appear—and most of the time, they comply.

Meanwhile, all of the aforementioned systems are often used simultaneously, depending on the circumstances.

These systems have been used for millennia for obvious reasons. For starters, most people fit easily into any of these systems. And historically, we had nothing else to go on. Only recently has science given us more precise ways of describing and explaining sex and gender.

Newer systems

Definition/system #4—size of gametes

These days, biologists use a definition that they can apply to most plants and animals. This system categorizes males as those who produce smaller gametes, and females as those who produce larger gametes. Gametes in humans are either eggs or sperms. By this definition, males include anybody who produces sperm and females include anybody who produces eggs.

This definition is extremely useful in biology. When we talk about plants, for example, it allows us to explain plant

Chapter 1—Intro and Definitions

reproduction by describing a plant (or a part of the plant) as either male or female based on a consistent definition.

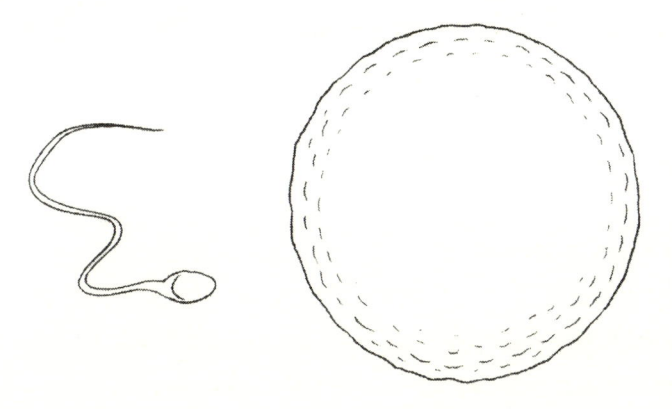

For humans, this definition also overlaps generally with the traditional definitions our ancestors used for both animals and humans where sex was determined by the role in procreation or by the presence or absence of a penis.

Definition/system #5—sex chromosomes

Another system for defining sex is based on sex chromosomes (or genotype). This relatively new system has now become pervasive in many sectors. This system categorizes human females as those who have two X chromosomes, and males as those who have one X chromosome and one Y chromosome. This same system applies to most mammals.

This system also has a huge overlap with all the above systems, and it works well most of the time. Currently, however, most people never actually have a genetic analysis to confirm their sex chromosome configuration, although DNA testing is becoming more and more common.

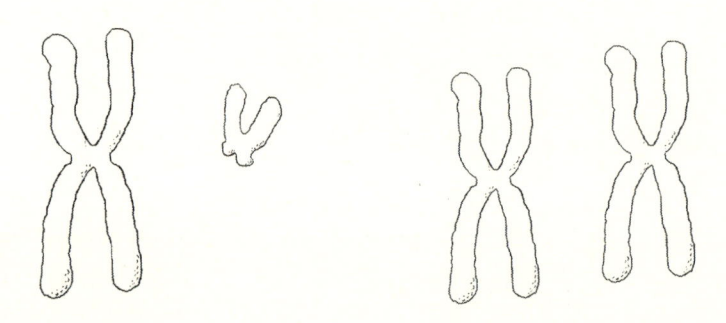

Shortcomings

All of these definitions are useful and are in common use, but they each have shortcomings because they all leave people out.

In fact, every one of these systems has exceptions that are non-controversial. For example, there are plenty of people who can neither get pregnant nor get somebody pregnant (system #1) or who produce neither eggs nor sperm (system #3). We still consider those people to be either male or female. So, each of these definitions is useful most of the time, but they are certainly not always precise or sufficiently inclusive. Even using chromosomes as a definitive system is faulty and in Chapter 6 I will list some of the situations where even this system is not adequate.

Definition/system #6—gender identity

A system that has emerged more recently is using our internal sense of gender identity as the most important factor in determining our gender. According to this system, gender resides in the brain and manifests as an innate conviction of one's own gender.

Summary

Other systems have been used in other cultures and at different times in history, but these are the most common systems in use today in Western societies.

To review, the six most common systems for determining gender are
1) *whether that person appears to have the characteristics of a person who is/was/will be able to become pregnant vs. a person who is/was/will be able to impregnate somebody,*
2) *whether or not that person has a penis vs. a vagina,*
3) *whether that person outwardly appears to be male vs. female based on presentation and visible traits,*
4) *whether that person contributes a larger gamete (e.g., egg) vs. a smaller gamete (e.g., sperm),*
5) *whether that person has XX sex chromosomes vs. XY sex chromosomes,*
6) *whether that person has an internal sense/awareness/conviction of being female vs. an internal sense/awareness/conviction of being male.*

Precision, exceptions, ideology

Going forward, I will use the terms "male" and "female", without being precise about which of the above systems I am using, unless it is relevant for the point being made. As an example, if I say, "most females have vaginas," it does not matter which of the above definitions of female I use because it is true in all 6 of the above systems.

To understand sex and gender, we need to understand the exceptions in the above systems. It is the rare situations that force us to clarify our meanings, and also that make our understanding of sex/gender complete. We have to look at situations where gender is more ambiguous or counterintuitive because these cases help us gain a more complete understanding of how our bodies and brains work.

The focus here is on information rather than ideology. Everybody has an ideology about gender, including me, but the purpose of this book is not to argue my ideology.

Defining "masculine" and "feminine"

In this book, I will be using the terms "masculine" and "feminine" frequently.

There are traits, or structures, or characteristics that can be more feminine or more masculine. These gendered traits include physical characteristics that emerge in our bodies. They also include structures and neural pathways in our brains, which form the basis for all of our personality traits. And since many brain structures have undergone sexual differentiation to become either masculine or feminine, numerous personality traits also show substantial sexual differentiation.

The words "feminine" and "masculine" can actually be much more precise (compared to "male" or "female") because these terms actually allow for more consistent definitions.

I will use the following definitions for the purpose of this book:

A feminine trait is any trait that is more common among females than males.

A masculine trait is any trait that is more common among males than females.

It doesn't much matter which definition of "male" or "female" we use because the exceptions in all six described systems are mathematically low.

Granted, we will not always have a full consensus on whether or not a certain trait is more common in males or females. Some people might even criticize the examples I am going to use to illustrate my upcoming points. I hope people who disagree with my examples will realize that I am not trying to prove any certain trait is more masculine or feminine but that I am just choosing examples that help illustrate how we develop gendered traits.

Before I move on from this subject, it is important to recognize the following: A large number of traits or characteristics may be either masculine or feminine, but every individual human has some masculine traits and some feminine traits. Most males have more masculine traits, but every male has at least some feminine traits. Most females have more feminine traits, but every female has at least some masculine traits. Every human is a mosaic of masculine and feminine traits.

Defining "divergency" or "divergent trait"

We are all mosaics of masculine and feminine traits. This is particularly true for any trait that resides in our brain. And since we are all a mosaic of masculine and feminine traits, you can say that we all have many "mismatches".

When I say "mismatch", I refer to either a masculine trait in a female or a feminine trait in a male. I call it a "mismatch" here for lack of a better word. I certainly am not implying that this is abnormal or pathological. These divergent traits are normal and complementary. It is actually how nature designed it to be. Humans have always benefited from having a wide variety of characteristics. And some of this variability comes in the form of the numerous masculinity/femininity "mismatches" that are present in all of us. However, in an effort to avoid the negative connotations associated with calling them "mismatches" I will refer to these as "divergent traits" or "divergencies".

Chapter 1—Intro and Definitions

Divergency: When I use the terms "divergency" or "divergent trait", I am referring to either a masculine trait in a female or a feminine trait in a male.

Sex vs. gender

In this book, I am using sex and gender mostly interchangeably. There are important discussions taking place about the different ways that sex and gender are defined and distinguished in our language, but that is not the point of this book. The point of this book is to educate the reader about how sexual differentiation develops in our bodies and our brains. Hopefully, this knowledge will give people a better basis for future discussions on the meaning of sex, gender, and how we manage sex and gender as a society (even though I won't be addressing any of those issues in this book).

Italic sections

Throughout this book, I will need to repeat these definitions and I will need to repeat other important mechanisms and processes that are recurrent themes. I will sometimes add summaries as well to reinforce important concepts. I am going to use italics to signal most of these repeats. They are important to understand, so I include them each time. I put them in italics so that if the reader feels that they already understand the idea(s) they can more easily skim through that paragraph.

21

r 2—Chromosomes part 1

Chromosomes

For most of the last century, sex chromosomes have been considered the bottom line when it comes to determining sex. It mostly works out because there is a very high correlation between the status of sex chromosomes and all the other possible systems of classifying sex/gender discussed in Chapter 1. However, there are some important situations where it doesn't work out.

Since our chromosomes carry all of our genetic information, they contain the information that guides every step of our development. We all have 46 chromosomes in each of our cells (except red blood cells and gametes), and they control how each cell grows, divides, differentiates, etc. Of these 46 chromosomes, two of them are sex chromosomes. Females typically carry two X chromosomes, while males typically carry an X chromosome and a Y chromosome.

Genes for sexual traits are found on all chromosomes

The genes for sexual traits and structures are not necessarily found on the sex chromosomes but are scattered throughout all of the chromosomes. For example, the genes that code for the development of a penis are not on the Y chromosome. They are

Chapter 2—Chromosomes part 1

found elsewhere—distributed among the other chromosomes. The genes that code for the development of a vagina are not all found on the X chromosome, although some of them might be. These genes are also spread out among all of the chromosomes.

The size and shape of the penis is determined by genes inherited from the mother and the father. The size of the breasts is determined by genes inherited from the father and the mother. The amount of chest hair is inherited from the mother just as much as the father. In other words, all of our sexual characteristics are inherited from both of our parents.

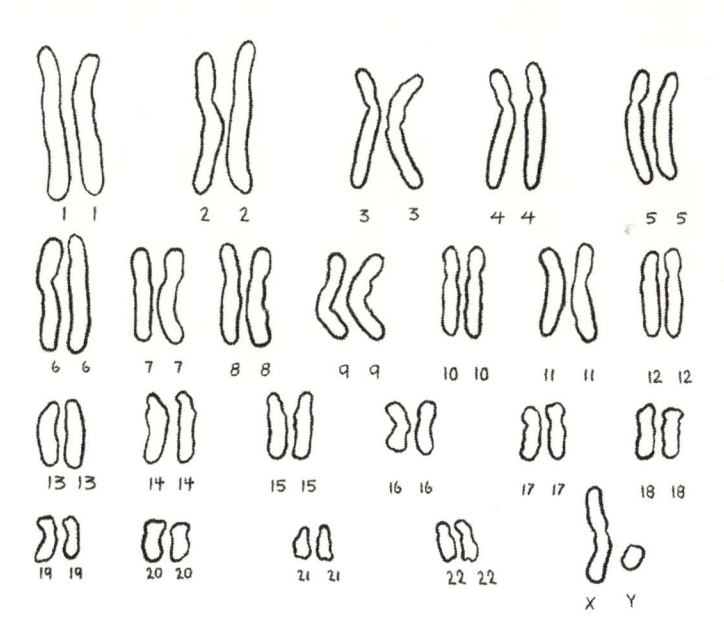

Everybody carries all of the female gene and most of the male genes

With some very, very minor exceptions, each of us has all the genes necessary to make both a complete male and a complete female. We all start out with that potential. It just comes down to turning those genes on or off (which is where the Y chromosome plays a role, as I will explain).

This universal potential has some interesting manifestations. Nipples and breasts help illustrate this.

Men's nipples serve no function, but all men have them. Furthermore, males have all the genetic information in their chromosomes to make perfectly functional breasts. These genes for breast development are usually suppressed in males. But sometimes, these genes are induced to develop by certain environmental toxins or other conditions. This can lead to breast development in males and can sometimes even lead to lactation.

By contrast, in females, the genes for breast development are routinely activated at puberty. The genes for lactation are further activated at the end of any pregnancy to enable breastfeeding. Still, both males and females have all of these genes, even if the males don't use them (other than to pass them on to their daughters).

Meanwhile, in males, the genes to make nipples are not suppressed and each male is born with two (or more) fairly useless nipples.

Like the genes for breast development, every male carries every gene for every female trait. These genes are spread out over 45 of their 46 chromosomes. (Y is the exception and carries no female genes.)

Females also carry the overwhelming majority of male genes. This includes the genes that code for development of the testes, the penis, and the genes that code for nearly every other sexual characteristic. These genes are located throughout the genome, spread out over all 46 of their chromosomes. In females, these male genes are present but not expressed (or expressed very weakly). Just like males and their breast genes, these male genes generally remain dormant in females, ready to be passed on to their sons.

The important exception is that females do not carry any genes that appear on the Y chromosome. The Y chromosome carries the only genes that are unique to males. But, this is one of the smallest chromosomes and only carries a very small number of genes.

As it turns out, the Y chromosome has one extremely important gene, the SRY gene. The rest of the genes on the Y chromosome are minor. The main thing the SRY gene does is to act very early in fetal development, around 6-8 weeks after conception, to turn on the testosterone-producing factory that leads to the development of testes.

The other minor genes on the Y chromosome seem to be related to sperm development and that impacts fertility. But when it comes to determining sex and gender, it is mostly the SRY gene, and it mainly just codes for turning on the switch. It does not code for any actual masculine structures and does not even code for testosterone. It simply triggers the unlocking of the initial masculinization pathways. Meanwhile, the genes for forming testes, scrotums, penises, prostates, etc., are all found on other chromosomes that both males and females share.

Summary: males carry all of the genes to make a complete female, and females carry almost all of the genes to make a complete male, with the exception of the SRY gene that flips the switch for starting masculinization and a few other minor genes also located on the Y chromosome.

What follows from this is that when a fetus is developing, whether it is male or female, each of its traits has a genetic template that has all the genes necessary to develop in either a feminine or a masculine direction. Since all of the genes are present, it comes down to how the genes are expressed. In the next chapter, we will talk about the processes that guide the sexual differentiation of gendered traits.

In Chapter 6 I will continue the discussion of chromosomes by first giving a basic overview of how our chromosomes work, including sex chromosomes. I will then give a brief overview of

Chapter 2—Chromosomes part 1

the different sex chromosome configurations that fall outside of the normal XX/XY binary. These uncommon chromosomal configurations highlight the shortcomings of basing sex or gender on sex chromosomes.

Chapter 3—Testosterone

Female and male bodies and brains

Certain essential features of our bodies, particularly our gonads and genitals, undergo sexual differentiation and become masculine or feminine. There are many other less-essential features of our bodies that also masculinize or feminize. And all along the way, different parts of our brains also differentiate into masculinized or feminized brain structures and apparatuses.

In the end, most males end up with male gonads (testicles) and male genitalia (penises). They also end up with predominantly masculine features but to varying degrees. Different individual males have different degrees of masculinization for any characteristic.

For example, the vast majority of adult males have much more growth in their larynx and develop deeper, more masculine voices. Most, but not all males develop some facial hair. Other masculine traits emerge somewhat less frequently such as prominent chest hair or male pattern baldness.

All these traits can emerge in males, but they have varied levels of expression in the population. In the examples above, the development of penises and testicles has an extremely high expression. This means that the percentage of males who develop penises and testis is very high. Lowering of the voice has a slightly lower expression but is still very high. Facial hair is expressed a bit lower than that but still fairly high. Chest hair has an even lower expression compared to these other masculine traits. These are all masculine traits that have various levels of expression.

We all have numerous physical traits that are masculinized or feminized. Besides facial and body hair patterns, we have body shape and size, including hip size, hand size, foot size, height, leg length, muscle mass, bone thickness, nipple size, breast development, etc. All of these traits have varied levels of expression in both males and females.

Definition: brain apparatuses

Humans also have numerous gendered traits that are based in our brains. We have personality traits, mental traits, neurological traits, instincts, and capacities that are either feminized or masculinized. These are made up of neural structures and neural pathways within the brain. Each has a unique architecture and unique interconnectivity that were impacted by sexual differentiation.

For the purpose of this discussion a "brain apparatus" is any particular brain structure comprising the neural tissues and connections that make up a particular trait. Each apparatus is made of tissues that can be located in different parts of the

brain but are interconnected and work together to perform a function.

Reminders:
—Any female will have some feminine personality traits and some masculine personality traits, but most females will have more feminine traits. Any male will have some masculine personality traits and some feminine personality traits, but most males will have more masculine traits.
—By my earlier definition, masculine traits are any trait that is more common in males and feminine traits are any trait that is more common in females.
—It doesn't matter what system I use to categorize males and females because the masculine/feminine traits I am referring to would predominate in males/females regardless of which of the six main systems I use to define male and female.

Sexual differentiation and testosterone

In order to understand how a human becomes male or female, we need a basic understanding of how each individual trait undergoes sexual differentiation. How does it become masculine or feminine or something in between?

That seems like a big task because we have thousands of gendered traits. The good news is that there is a basic path of differentiation that these traits commonly follow, which we can easily learn.

Some experts have argued that we are all female by default—that we all start out female, but some of us differentiate into males. It is true that if the SRY gene doesn't activate the testosterone factory, then all traits will proceed in a feminine direction. However, I prefer to assert that we all start as undifferentiated humans, and those of us who have a functional SRY gene will proceed in a male direction, and those who don't will proceed in a female direction.

Some might say that it is the testosterone that makes the man. Testosterone is indeed the factor most responsible for the masculinization that happens to male bodies and brains. If you need a simple narrative, then that one is useful. In fact, I am going to use that simplification throughout the coming explanations. Everything is actually more complicated than that, but most of us are going to understand it better if we think of it as being "all about the testosterone".

(For those who need more precision: What makes it more complicated is that testosterone is an androgen, and there are other less important androgens. Furthermore, all males and all females need exposure to some amount of all the sex hormones, including the male sex hormones such as testosterone and the female hormones such as estrogen and progesterone. Furthermore, testosterone in lower amounts is essential to normal female development, and estrogen is actually made from testosterone. Our development is very concise and complicated and involves precise timing and a balance/interplay of all these hormones. So from now on, every time I say that "testosterone" does something,

what I really mean to say is "testosterone, along with other androgens working in a precise balance with other hormones.")

When I say:

TESTOSTERONE——>EFFECT

I really mean:

TESTOSTERONE PLUS OTHER ANDROGENS WORKING IN A PRECISE BALANCE WITH OTHER HORMONES——>EFFECT

High testosterone levels lead to masculinization

Simply put, the overarching theme is that if you expose developing tissues to large amounts of testosterone at a critical stage of development, it will masculinize. And if you don't, it will feminize. (This only applies to tissues that undergo sexual differentiation.)

Let me repeat that point to really emphasize it. If you expose a developing tissue to large amounts of testosterone at a critical stage of development, it will masculinize. And if you don't, it will feminize.

Let's break down that statement a little bit and explore a few of the caveats.

Tissues that undergo sexual differentiation can be in our bodies or in our brains. This includes the fetal tissues that will later grow into gonads or genitals. This includes the tissues in our larynx that will grow at puberty and will determine the pitch of our voices. This includes our bones and muscles when they decide how big to grow at puberty. This includes all the hair follicles on our chins that decide whether or not to produce thick hairs to make a beard vs. peach fuzz. This includes the hair follicles at our hairlines that decide whether or not to recede, etc.

This also includes thousands of very precise regions of the brain that either masculinize or feminize (or something in between).

The brain—a patchwork of feminine and masculine structures and traits

There is no such thing as a brain that is globally female or globally male. The differences that have been found are only differences in averages. For example, there are several structures in the brain where the average relative size might be different in males vs. females. However, there is a huge overlap to the point that you cannot determine the gender of any brain by any single characteristic.

So, when it comes to the study of measurable structures in the brain, everybody has a mix of masculine and feminine structures. What we have learned about personality traits also applies to these measurable brain structures: Most males have more masculine structures, but all males have both masculine and feminine structures. Most females have more feminine structures, but all females have both masculine and feminine structures.

Summary:
—There are no universal differences between female and male brains; there are just trends.
—Every brain has some masculine and some feminine structures.
—Every human has some masculine and some feminine personality traits.

Brain apparatuses can be localized but are mostly diffuse

Our personalities are based in our brain. However, the location of each personality trait is within a diffuse structure that can not be discretely measured with imaging. Each trait arises from different neural tissues and neural pathways distributed in different regions of the brain, each with its own unique architecture and interconnectivity. Although you can't discretely map out the location, you can often get a rough idea of which areas of the brain might be involved using modern scanning techniques such as PET scans, FMRI scans, or magnetoencephalography (MEG). Perhaps future technologies will allow for very precise mapping.

Even though we can't identify a concise, measurable brain structure for most personality traits or brain functions, it is still correct to say that there is an apparatus in the brain for every particular trait or function. Each brain function and personality trait is spread out through different brain regions and involves different brain structures. A widely described example is the language apparatus. This apparatus includes some localized structures (including the Wernicke's and Broca's regions of the cortex) but also involves some other more diffuse regions of the brain.

Critical periods

Before we go any farther, we have to really understand what a critical period is.

A "critical period" is a discrete period during the development of any trait when it is extremely susceptible to signals from the environment. That trait's final pathway will be established during this short period of time.

When it comes to sexual differentiation, a critical period is basically a fork in the road, where a particular trait will proceed in either a feminine or a masculine direction, depending on the local environment during that stage. Critical periods are "now or never" moments during development.

In the research community, critical periods have mostly been discussed when referring to periods of time where sensory input is crucial to development.

One example of this is the development of the brain's ability to process sound, which is needed for hearing. A child must be exposed to sounds during the critical period of development in order to develop hearing. If a child is not exposed to sounds in the early years of life, as could happen if they have deafness or a mechanical issue in their middle ear, their brains will not be able to process sound. Even if the ear problem is diagnosed and repaired later, that person will never develop the ability to process sounds if the critical period has already passed.

There has been a lot of research about critical periods for things like binocular vision, hearing, and language acquisition.

Meanwhile, critical periods of sexual differentiation have not yet gotten as much discussion. However, they really should. Critical periods are common and are almost universal in all aspects of the development of both our brains and our bodies. This is true for traits that are sexually differentiated as well as traits that are not sexually differentiated.

"Critical periods" is one of the most important concepts that we are going to discuss in the next chapters. Aspects of these processes are well understood by embryologists, developmental biologists, neurologist, neurobiologists, endocrinologists, etc., but I haven't found a good synthesis of all of these disciplines when it comes to sexual differentiation. What you are reading is my attempt at an overarching narrative.

Some researchers distinguish "critical periods" from "sensitive periods", where a "sensitive period" is a window where the tissue is highly sensitive to environmental conditions, and a "critical period" is a special kind of "sensitive period" where normal development absolutely depends on certain environmental inputs. I will not be making this distinction in the following discussion and will refer to all these types of situations as "critical periods".

Why it is so critical to understand critical periods

The development of a human body follows an exact recipe of events that happen in a very precise order. You don't want to grow fingers before you grow arms, or the fingers might end up on the wrong part of your arm. You need a brain stem before you can grow a cerebral cortex. Precision and timing are key. Critical periods are a strategy biology uses to exert control over the building process. The mechanism helps ensure a step-by-step sequence following a set of blueprints provided by the DNA.

Let's take the critical period for finger development. In this case, there is a critical period that starts at the beginning of finger growth. It happens early in the pregnancy.

During the critical period, finger growth starts as just a bud. Then it starts to acquire its architecture and the tissues develop into bones, muscle, etc., following a step-by-step process. If the process is interrupted for some reason before the finger has its form and architecture, it will never grow into a finger. The finger can only grow if it acquires its basic structure during this early period. In this case, the critical period is during an early stage when its tissues are differentiating and its architecture is being established.

During this critical period of finger formation, it is more sensitive to its local environment. It is more sensitive to hormones, to toxins, and to the availability of nutrients. Environmental factors will always have an exaggerated impact during critical periods.

To put it simply, one day the fetus has no fingers, then it will start growing finger buds. While the hand and fingers are acquiring their architecture, they will be very sensitive to environmental signals that will help them grow correctly—this is the critical period.

Soon it will be evident that their architecture has been fully established, including fingernails, and the fingers will have grown to a length that is proportional to their small hands. This will mark the end of the critical period for finger growth, even though we know that the fingers will continue to grow longer and larger until adulthood.

If the process is interrupted for some reason during a critical period while the architecture is being established, the child can be born without a finger or limb, or with a limb that is only partially formed. Once that happens, it is too late. That limb can never fully develop. A sad example of this is thalidomide, which was prescribed in Europe in the late 1950s to treat morning sickness. This medicine turned out to be toxic to the fetus and caused numerous problems, including interrupted limb development. Even though this medication was safe for a fully formed adult, if it was present in the fetus during critical periods of limb development, it interrupted the limb formation process.

Meanwhile, every human body part, every human characteristic, every personality trait, every mental trait, etc., passes through one or more critical periods. The tissue responsible for that trait has to be activated (or suppressed) at exactly the right time. The expression of the relevant genes has to be turned on and off at exactly the right time. During those critical periods, anything that impacts the development of that trait will have an exponentially greater impact on the final outcome.

Sex hormones and critical periods

For developing tissues that undergo sexual differentiation, sex hormone levels are vitally important to the outcome, especially during critical periods. It turns out that testosterone plays a determinative role in the sexual differentiation of every gendered structure in male bodies and brains, particularly during critical periods. During critical periods, testosterone impacts which genes get turned on and which genes get turned off in these developing tissues.

Embryologists refer to this process as "activation". Testosterone "activates" tissue masculinization. If there are high enough levels of testosterone during a critical period of development for that tissue, then that tissue will be activated to develop in a masculine direction. Otherwise, it will proceed in a feminine direction.

So, these critical periods for sexual differentiation are a very discrete time period when the levels of testosterone will determine the final outcome for that trait. Whether it be a sex organ, a physical trait, or a personality trait, the levels of testosterone will play a determining role during critical periods. After each critical period ends, the fork in the road has been passed. The direction is now fixed. The general outcome for that trait is now inevitable.

Sometimes, that critical period is very short and might last only hours or days. Other critical periods could be months or even years. (Caveat: there are certain traits that are impacted by testosterone levels throughout life. These same traits also have early critical periods that require high levels of testosterone to set them on a masculine course, but even at later times, testosterone might continue to act on that tissue.)

Remember, anytime I talk about testosterone acting on a tissue, it is acting in balance with all the other sex hormones including other androgens, female hormones, and other hormones that are not sex hormones. Furthermore, even though high levels of testosterone are required to activate masculine development, the presence of lower levels of testosterone is also important for feminine development.

Sexual differentiation

Here is a summary of the process as it applies to sexual differentiation. This process happens over and over again for every gendered structure or trait:

1) We have a tissue in the body or brain that has the potential to eventually develop into a structure that is either masculine, or feminine, or somewhere in-between.
2) At a certain stage in development, with precise timing, that tissue will begin to develop and differentiate.
3) At a critical period during development (usually early), that tissue is highly sensitive to the presence or absence of high testosterone levels (working in balance with other hormones). The presence of higher testosterone levels during this period will set that tissue (or trait) on a path toward masculine development. The absence of high testosterone levels will set that tissue (or trait) on a path toward feminine development. There can also be intermediate outcomes. In fact, when it comes to brain traits, most of the outcomes are intermediate.
4) Other environmental factors will also play an outsized role during critical periods.
5) The blueprint (coded in our genes) guides and individualizes each step in the process. (Note that genes will also impact the final outcome through other mechanisms that are independent of testosterone.)

Genetic factors

Let's further discuss the interplay between genetic factors and hormones using the penis and clitoris as an example.

The penis and the clitoris develop from the same tissue. Basically, a penis is an enlarged clitoris containing an extension of the urethra. In every infant, the levels of testosterone determine whether the same tissue will grow and develop into a penis, or a clitoris, or something in-between.

Testosterone levels do have a substantial impact on the final size, but when it comes to penis size, other genetic factors also play a role. Each male child inherits genes from both the mother and the father that impact the final size of the penis. So the final size of any male's penis is impacted by his mother's genes, his father's genes, and the amount of testosterone present during certain critical stages. Other hormones and other factors such as nutrition, and environmental toxins, also play a role.

Other sexually differentiated traits work in a similar way.

Sensitivity to testosterone

For any tissue in the brain or body that has sexual differentiation, there may be some variation in the sensitivity of that tissue to the presence of testosterone. This would be dictated by their genes. Some fetuses might inherit genes for a certain tissue that are highly sensitive to testosterone levels, and some might have genes that are less sensitive (or resistant). If a tissue is not sensitive to testosterone, then the presence of testosterone will not make a difference one way or the other. If the tissue is highly sensitive to the presence of testosterone, then the amount of testosterone will have a bigger impact on the final outcome. This is another way that inherited genes can impact the extent of sexual differentiation of certain traits.

Interactions with the environment and epigenetics

There are environmental factors that also impact development. It is useful to discuss the prenatal environment and the postnatal environment separately. Postnatally (after birth), the "environment" usually refers to a child's surroundings, including the air, food, and water, but also includes how their parents raise them and other social and cultural influences.

However, much of development happens prenatally. The prenatal "environment" is essentially the fluids that the tissues are soaking in. This includes whatever is circulating in the fetal blood, in the cerebrospinal fluid, or in the amniotic fluid. These circulating factors include nutrients, hormones, antibodies, toxins, and whatever else happens to be present at the time.

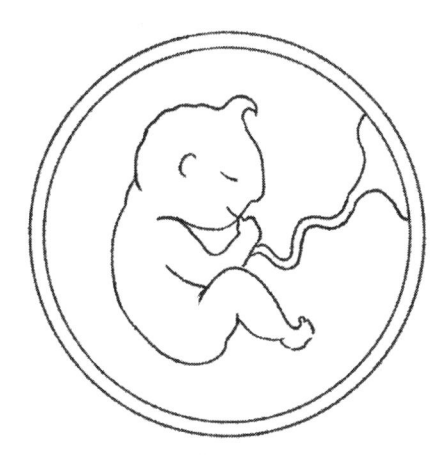

The prenatal environment is built as a (mostly) cooperative effort between the mother and the fetus. Circulating hormones, including testosterone, are part of that cooperative effort. Many of the hormones that are present are supplied by the fetus itself. Others are produced by the placenta. Yet others are provided by the mother and are passed from her bloodstream to the fetal environment by way of the placenta.

There are many environmental factors that are supplied entirely by the mother, starting with all the basic building blocks and nutrition. Due to life circumstances, there may be some variability in the availability of these resources to the fetus from the mother. The mother might be going through hunger or an illness that impacts her ability to absorb nutrients. She will be navigating her own life and having her own reactions, including intermittent stress. Maternal stress will impact the hormones she produces, as well as the amount of nutrients that she can pass on to the fetal environment.

The mother can also be exposed to viruses, bacteria, or allergens that lead to immune or allergic responses. This leads to antibody production. Some of the resulting antibodies are passed on to the fetus and will impact development. Furthermore, the mother can have autoimmune reactions or even immune reactions against the fetus, and those immune factors can sometimes cross the placenta and impact development.

Last of all, toxins can be present in the mother's bloodstream and can be passed on to the fetal environment. This is becoming more and more relevant over the past few decades with increasing levels of microplastics and pesticides in our food and water supplies. Many of these are endocrine disruptors, meaning that they interact with our hormones in ways that impact sexual development.

Other toxins might be intentionally ingested by the mother, such as medications or drugs or alcohol. These can also impact the developing fetus.

When any non-genetic factor impacts gene expression it is called "epigenetics". Thus, in fetal development, epigenetic factors refer to any non-DNA factors that impact gene expression, including environmental factors. Epigenetic factors regulate when certain genes are turned on and off. This, in turn, affects how that tissue develops and the final outcome of that trait.

During fetal development, the environmental factors that have an epigenetic impact are acting very locally, at the level of each cell. Some of these epigenetic factors are endogenous or created by the fetus, and others are exogenous or come from outside the fetus by way of the mother. Testosterone levels are an important factor that have an epigenetic influence on gene expression for each cell.

Epigenetics research aims to explain the mechanisms for gene activation and suppression. One day, this research will hopefully explain more precisely the mechanisms of how testosterone impacts gene expression during critical periods and why those changes persist once the critical period ends.

Let's go back to the example of the penis. In a male child, the penis will start developing if development is triggered during early pregnancy. The final size will be determined by
1) the genes inherited from the mother and father and coded by DNA in the chromosomes,
2) the amount of testosterone present during critical periods,
3) the overall sensitivity of that tissue to the effects of testosterone,
4) other epigenetic factors which are the environmental conditions present when that tissue is developing, including anything the mother might have circulating in her bloodstream that can cross into the uterus (toxins, hormones, antibodies, nutrients, etc.).

All of the above is true for any sexually differentiated physical structures in our body. It is also true for the thousands of regions of our brain that are subject to sexual differentiation.

Our brain structures develop in a precise, time-dependent fashion. They are built layer by layer. As different layers are added to the brain, each new tissue will have a critical period where it can be subject to levels of testosterone and other hormones, as well as nutrients, toxins, antibodies, and other environmental factors.

Critical period for gonad development

In males, the SRY gene on the Y chromosome activates the testosterone factory very early. In fact, by week 7, the SRY gene in a male fetus is stimulating testosterone production that leads to development of testicles, which leads to even more testosterone production.

The action of the SRY gene will also suppress the development of female organs and cause some proto-female tissues to be deactivated and reabsorbed. If the SRY gene is not present, then ovaries will develop instead of testes. By week 8, these early gonads can be identified in fetal tissue.

Thus there is a "critical period" around week 6 or 7 that determines whether you develop testes vs. ovaries. That is the fork in the road. The SRY gene and its products determine which fork is taken. After the critical period ends somewhere before the end of week 8, there is no going back, and the SRY gene is no longer relevant to gonad development.

Critical period for genital development

Differentiation of the genitalia follows immediately after gonad development, occurring from weeks 8–12. Critical periods for the development of penis/scrotum vs. clitoris/vagina happen in this time period. The levels of testosterone are going to be determinative during those weeks.

After this critical period ends, no matter what happens, there is no going back. There will still be further growth, but the direction has been set. If that person has a penis at week 12, they will be born with a penis. If they have a vagina at week 12, they will be born with a vagina. Testosterone levels and other hormones might continue to play a smaller role in the final outcome, but the critical period is over. Genital differentiation has passed the fork in the road.

Critical periods at puberty

Overall, many body parts and traits have critical periods for sexual differentiation. Some of them are early, such as gonad and genitalia development. Some of them happen later during pregnancy/gestation. Others happen much later. I am going to jump way ahead and discuss critical periods that happen during adolescence but are critical periods just the same.

At puberty, a surge in hormones triggers certain body parts to change and develop secondary sexual characteristics. These changes impact many different tissues in both sexes. Hair starts growing on the body in different ways. Bone and muscle growth occur in different patterns based on inherited genes as well as the cocktail of hormones and other chemicals that are present.

In females, the ovaries and uterus work together to start ovulation and menstruation while the breasts develop and enlarge. In males, the larynx is induced to enlarge, causing a voice deepening. In both sexes pubic and axillary hairs appear early, but differing body and facial hair patterns will eventually emerge in later puberty.

All of these changes involve critical periods, where levels of sex hormones at specific stages play a role in activating or reinforcing a pathway toward masculine or feminine development.

Hormone surges instigate puberty, and during critical periods, these hormones determine whether puberty will proceed in a masculine or in a feminine direction. It does not take long before the path is set in a way that is basically irreversible.

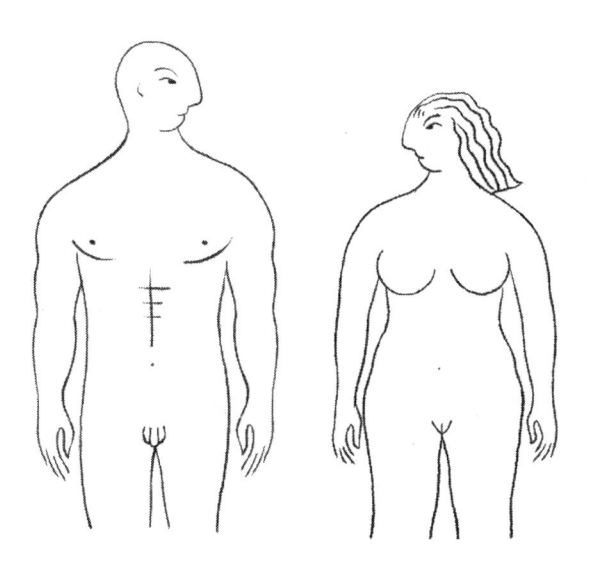

Divergent traits

Remember when I use the term "divergency" or "divergent trait", I am referring to either a masculine trait in a female or a feminine trait in a male.

The two most important causes of divergent traits are (1) inherited genes and (2) the timing of critical periods.

1) Inherited genes can clearly cause divergencies. We each inherit individual tendencies for each particular trait. A good example is facial hair. There are plenty of males out there with plenty of testosterone but don't grow much facial hair. This trait is feminized in these males because of the genes they inherited.

Inheritance can play a role in the level of sexual differentiation of almost any trait, and it will lead to different outcomes from one trait to the next within the same individual. This is because each of these traits is coded for by different genes. We have all inherited some templates that are inherently more masculine and others that are inherently more feminine. Inheritance also determines how sensitive or resistant that trait will be to hormone-induced differentiation.

2) A second big reason for the varied outcome is the variable timing of critical periods. Critical periods are happening at different times throughout our development. Meanwhile, hormone levels and other environmental factors can change from day to day. Thus, this variable timing of critical periods leads to more varied outcomes of these traits due to exposure to varying levels of testosterone and other environmental factors.

This is especially relevant in brain apparatuses and personality traits. The brain develops in multiple stages and each brain structure will have different critical periods where it is highly sensitive to testosterone levels. These critical periods for brain structures happen later than the critical periods for gonad and genital development. This difference in timing facilitates the development of divergent traits.

For example, there might be a brain apparatus with a critical period at week 14 of gestation—let's call this one apparatus A. There might be another with a critical period at week 16—lets call this one apparatus B. The testosterone level might be very different at week 16 compared to week 14. Therefore, those tissues of apparatus A will have a different level of testosterone exposure during its critical period compared to apparatus B. There might also be other environmental differences at week 16 compared to week 14. There might be different levels of other important hormones or different levels of toxins, or different levels of available nutrients. These timing differences can lead these two tissues within the same fetus to end up with different levels of masculinization or feminization. And remember that both of these traits have critical periods that are much later than gonad and genital development. So, each trait can differentiate to a different extent and even in different directions, and each may or may not diverge compared to the gonad/genital configuration.

Brain development also tends to favor intermediate outcomes. This is because brain structures tend to have a less binary response to testosterone compared to the more binary response of gonad and genitalia development. Brain traits are mostly determined by our genes, and the sexual differentiation that does occur tends to have a lower overall impact. Just the same, there are numerous brain traits with an observable average difference between males and females and we will discuss this more in Chapter 4.

In the end, most personality traits are neither fully feminine nor fully masculine but somewhere on a gradient between the two extremes. This is the result of an interplay between the inherited genes, their sensitivity to hormones, and the hormone levels during critical periods. This gradient adds to the variety within the mosaic of our personality traits. We all end up as a patchwork of some highly feminine traits, some highly masculine traits, some moderately feminine traits, some moderately masculine traits, and some traits that fall close to the middle.

Nature vs. nurture

There are a lot of debates about "Nature vs. Nurture". This involves trying to determine whether a certain trait is due to inherited genes vs. how the person was raised. However, anybody who is familiar with development will tell you that these "nature vs. nurture" debates don't always make sense. Everything is nature AND nurture.

This is how "nature and nurture" work:

1) Every trait has a template that is determined by a set of genes that are coded for by DNA. These genes are inherited—thus the template is inherited.

2) During fetal development, each trait template is subsequently modified by the local environment (epigenetics), meaning the fetal environment plays a role in how that trait develops.

By the time we are born, our traits are already impacted by both nature and nurture, with the "nature" being the DNA code and the "nurture" being the fetal environment.

3) After birth, all of these traits are subsequently acted upon by the environment, and in this case, the environment includes the food, water, and air, not to mention toxins and diseases, but also includes the social environment, the parenting, the culture. This is part of the ongoing "nurture" of that trait.

4) For any particular trait, it might be difficult to determine how big a role nature and nurture each play. This applies particularly to gendered personality traits.

There have been ongoing debates about "nature vs. nurture" for gendered personality traits. Some argue that there is very little difference between male and female brains at birth and that most difference in traits is related to the social and cultural environment they are raised in, as well as the numerous ways their caregivers encourage conformity with their gender. Some even take a more extreme viewpoint and assert that there are no meaningful differences between male and female brains.

Even though there is no universal "female brain" or "male brain," there is compelling evidence that there are numerous features including structures, functions, and abilities, that have masculinization or feminization. This means that you will find a difference between the average of males and the average of females within the population for any of these structures or traits.

This does NOT mean that all females have a feminine version of any particular trait or that all males have a masculine version of that trait. There are virtually no brain traits that are segregated that way. You can not take any particular brain trait in any particular individual and assume it is feminized or masculinized just because that individual is male or female.

Overall, every trait develops based on the these five factors:
1) *inherited DNA code*
2) *fetal environment, including hormones, especially testosterone, when sexual differentiation is involved*
3) *the physical environment the child is raised in, including air, food, water, toxins, pathogens, etc.*
4) *the social environment the child is raised in, including parenting, culture, socialization, siblings, peers, education, etc.*
5) *the internal environment provided by the child's own hormones, etc., throughout their life*

The relative importance of each of these 5 factors for most traits has not always been firmly established, and it will continue to be subject to much debate.

I hope that disagreements about the relative importance of each factor won't get in the way of understanding the mechanism or critical periods, even if you don't agree with the examples presented here.

Personality traits are based on brain structures and their connections

It is important to repeat here that our personalities are based in our brains. Regardless of when in life we develop any particular personality trait, it is based in the brain. All of our preferences, interests, affinities, and tastes are based in our brains. As I discussed above, any personality trait we have is due to a brain apparatus that drives that trait. That includes everything, from what fascinates us, to what bores us, to what annoys us, to what makes us happy or sad or angry. It drives where we excel, and where we fail.

As I explained above, each trait is based on an apparatus that is spread throughout various distinct regions of the brain. Even though each apparatus is distinct, each one also shares structures with other apparatuses. So, in the end, the trait is a sum of all the parts that make up that trait. For example, a personality trait is the product of all the relevant brain structures involved, in the diverse brain regions involved, with the various interconnections involved. Since each trait can involve different regions, and since each of those different regions can develop at different times, then any trait might be subject to more than one critical period. This will add even further diversity to the possible outcomes.

Sex hormones can also lead to temporary changes in the brain

So far, we have discussed personality traits in the brain and how they develop, and how hormone levels during critical periods have a lifetime impact. The changes made during these critical periods are mostly permanent.

However, we know that hormones also have temporary effects on our personalities that are only manifested while the hormones are present.

The most obvious examples are some of the personality changes that come during puberty in both males and females.

By adolescence, our brains have already undergone substantial sexual differentiation, most of which happened prenatally. The massive hormone surges that occur during puberty induce further brain-based personality changes. Some of these changes are permanent, but some of the personality changes we see with puberty (and beyond) depend on the ongoing presence of the sex hormones. As these hormone levels decrease after puberty, some of the drives and behaviors that are so intense during adolescence become more moderate. Then as life proceeds, these hormones decrease even further with notable personality impacts. So, even though these traits might be fairly long-term, not all of them are permanent.

These hormone-induced personality changes have been fairly easy to study because they happen at an age when it is easier to measure hormones. We also have lots of situations where sex hormones have to be administered to individuals, and we can observe the impacts. There are other situations where hormones are abruptly interrupted, such as when the gonads are surgically removed, and these cases also allow observation of the personality effects.

Chapter 4—Apps

Smartphone Apps

Let's think of every human as if they were a smartphone. Every smartphone has its physical components (hardware), an operating system, and a bunch of apps.

Just like smartphones, humans have physical components, an operating system and apps. The physical components include our bodies and the physical structures of our brains. The operating system and apps are made from the architecture of our brains' neural pathways and their interconnections.

The analogy isn't exact, but it is useful for understanding how our brains and bodies work, especially when it comes to gender and sex.

"App" is an abbreviation for "application", but it also can be used as an abbreviation for "apparatus". This is convenient, because when I refer to any "human app," I am referring to an apparatus made up of neural tissues in different parts of the brain, with its own unique architecture and interconnectivity, along with the body parts that it interacts with.

Example #1—language

Let's look at some specific examples. This first example is a trait that has limited sexual differentiation, but it will help illustrate the analogy.

On your smartphone you might have a language-learning app like Duolingo. This app uses different parts of your phone and its operating system to perform a particular function, in this case, helping you learn a language. Various hardware and software structures of your smartphone work together to make this app functional. It uses computing, graphics, sound processing, etc. It uses the screen to display the information in a way that makes sense to the user. It uses the phone's speakers to pronounce the words and phrases so you can hear them. It uses the touchpad for input of instructions and uses the microphone for vocal input. It uses memory to store the information. This app is downloaded as a small program and only takes a small amount of storage when not in use. When it is in use, it temporarily takes over different parts of your smartphone and coordinates them to do the job at hand.

Now, let's look at a human version. You could say that we have a built-in "human app" for language learning. The app is located in your brain in precise regions, but when it is in use, it pulls in other parts of the brain and body. It uses the ears and auditory nerves to receive the sound signals. It uses the sensory part of the brain to process these sound signals and uses the language part of the brain to make sense of them. It uses the memory part of the brain to store the information so that we can use it later. It uses the motor parts of the brain to send signals to the muscles of the tongues, lips, diaphragm, and larynx so that we can produce speech.

More succinctly, this app directs diverse regions of our brains to interact with parts of our bodies in order to take in information, process it, store it, and use it to send out communication when needed.

Looking at it this way, you can see that we have numerous "human apps"—tens of thousands. We have an app for learning languages. We have an app for making conversation. We have an app for dancing. We have an app for throwing a baseball. We have an app for breathing. We have an app for deciding when to eat. We have an app for appreciating art and another that determines our tastes in music. We have apps that determine what fascinates us, bores us, angers us, attracts us, and more. We have apps for all of our behaviors, our preferences, our personality traits, etc.

Chapter 4—Apps

In summary, each app involves a set of diverse regions of the brain which interact in order to take in information, process it, and then direct our response to that information by sending out signals to other parts of the brain and body. Any particular app might take advantage of certain sensory organs in our bodies (e.g., ears, eyes) and might then execute an action that will be carried out by certain muscles in our bodies (e.g., tongue muscles, arm muscles) that control our movements, speech or behaviors.

An important difference between our "human apps", compared to smartphone apps is that every individual human version is unique. An app like Duolingo has very little variation. It might vary from one update to another, and it might work differently on Android vs. iPhone, but these smartphone apps have only limited variability. In the case of our human apps, each app is custom-made as our brain develops. Even though one person's language learning app will work similarly to anybody else's, there will also be lots of variability from person to person. It might end up being more effective in some people than others. It might have different strengths or weaknesses. Each person will have an optimal way to use their app that will be specific to them.

These differences arise from a number of factors, such as inherited genes, prenatal environment, critical periods, and postnatal environment (education, etc.). You will probably recognize that these factors are the same factors that I described above when talking about sexual differentiation. Like many brain traits, this particular trait (language learning) has some sexual differentiation, so hormones levels also play a role.

This overall pattern of development works this way for the vast majority of our developmental pathways.

Apps with sexual/gender differentiation

A huge number of our "human apps" come in versions that are clearly gendered. Each person will have a version of each of these apps that is either more masculine or more feminine or in-between. As stated above, an "app" (apparatus) refers to areas in the brain that work together to perform a particular function. All of these brain areas have physical (neural) structures and unique interconnections and pathways that impact how they work. In a developing fetus, the development of these brain structures is influenced by testosterone levels during any critical periods for that particular structure. The two most important factors that will determine how much that app will masculinize or feminize are inheritance and testosterone exposure during critical periods. Overall, each infant will end up with a unique set of apps that will include some that are more feminine and some that are more masculine, giving them their unique mosaic of apps.

Let's look at several examples of traits that are typical of "human apps" that might have masculinized or feminized versions.

Important disclaimers

Disclaimer 1: generalizing

Chapter 4—Apps

In the examples below, I am using the knowledge of how things work generally, and applying it to specific examples. The basic pattern is very well established. It involves stepwise differentiation and critical periods where developmental pathways encounter forks in the path, and once a pathway is determined there is no going back. This pattern is so common that it is almost universal throughout our development. However, the specific examples I use have not all been studied in this way. With that in mind, I am choosing highly probable, yet unproven examples, because they help illustrate how the process typically works. Since most developmental processes work this way, it is an excellent starting point. Hopefully, as our knowledge increases with further research, these ideas will either be reinforced, modified, or even replaced to reflect our updated understanding. Understanding the basic processes will be a great starting point for understanding any nuances and exceptions that will later be discovered.

Disclaimer 2: stereotypes

Stereotypes are useful here because they are intuitive, and they help illustrate how the process works. Understanding them also helps lead to a discussion of the exceptions to these stereotypes. I acknowledge that there are tons of ways that stereotypes can be misused, misleading, or even completely false. However, I am trying to use examples of gendered traits that are pretty obvious to most observers in order to increase understanding of gender development. I am certainly not trying to predict the destiny of any particular male or female, and I hope that remains clear as we proceed. A generalization can be useful for understanding a category of humans but is often terrible for understanding an individual. I am hoping that my upcoming discussion will end up supporting that idea.

Examples #2, 3, 4—some well-studied apps

Let's continue with more examples.

Example #2—face gazing
There is an app that is evident in infancy that determines interest in looking at faces. On average, female babies will spend more time than male babies looking at the faces of their caregivers. So you can say that a feminized version of this app leads to an infant with a higher interest in looking at faces.

Example #3—mentally rotating objects
Another app determines a person's ability to mentally rotate a 3D object. An example of this could be to show a subject several images of hands pictured at different angles and then time how fast they can determine which ones are right hands and which are left hands. From an early age, boys typically do better on these kinds of tests. So, you can say that a mascu-

linized version of this app leads to a child who can more easily mentally rotate 3D objects.

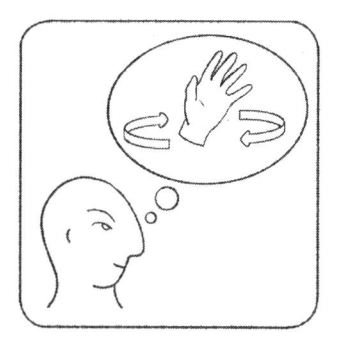

Example #4—recalling locations of hidden objects

Yet another app determines a person's ability to remember the hiding places of a number of hidden objects. In this case, a feminized version of this app leads to a capacity to remember the locations of a higher number of hidden objects.

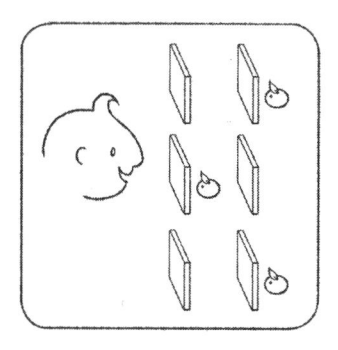

I mention these three traits because they are very well studied and appear early, before parenting or culture are likely to impact the outcome. Just the same, for all the traits that I describe, there are always a few experts who believe the sexual differences are completely due to differences in how caregivers treat male infants vs. female infants. In this case, I think the evidence is pretty strong that prenatal testosterone levels play a role.

But even if you don't agree with me on the examples I provide, please stick with me because I am trying to illustrate how differentiation works. There are certainly gendered traits that are, indeed, 100% dictated by culture and upbringing. On the other hand, there are many important traits that have a clear, neurologically-based sexual differentiation. Mostly, it is "nature AND nurture".

Example #5—innate interest in automobile mechanics

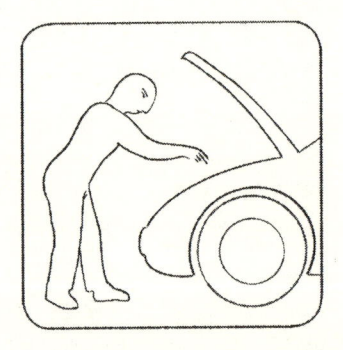

There is an app that determines how much interest a person will have for automobile mechanics. This "human app" includes several brain structures that interact to dictate that person's innate level of fascination with the workings of cars.

If we look at the population of adult males and the population of adult females, you will observe that there are more males than females who are highly interested in auto mechanics. Therefore, a person with a masculinized version of this "human app" will have a higher interest in auto mechanics, and a person with a feminized version will have a lower interest in auto mechanics (based on the definitions of masculine and feminine given earlier).

Note that there are lots of exceptions. There are plenty of men who have no interest in auto mechanics and plenty of women who have a lot of interest in auto mechanics. There are also plenty of people of both sexes who fall in between and have only a limited interest in auto mechanics.

So how does this happen? How does this app become more masculine or feminine? For the purpose of this illustration, we are going to assume that this app is masculine or feminine in any individual based on more than just social learning. We will assume that inheritance and testosterone levels also play a role. This is probably the case for this app but has not been proven.

Here is the basic process:

1) The fetus starts out with inherited DNA that carries the genetic code for all of their traits. This includes genes that will code for the structures that make up this particular app (interest in auto mechanics).

2) Even before sexual differentiation, these genes will code for a specific template for this app. This inherited template might

happen to be more feminine or more masculine, independent of the sex of the fetus. However, this is only a starting point.

3) As the different regions of the brain develop, the structures that make up this app will each pass through one or more critical periods. During these critical periods, they are sensitive to the levels of hormones. If there are high levels of testosterone during critical periods, then the structure(s) will become more masculine. If there are lower levels of testosterone, then the structure(s) will become more feminine.

4) Meanwhile, the genes that code for these structures will also dictate how big of an effect the testosterone will have (sensitivity vs. resistance to the effects of testosterone).

5) It is important to note that both male and female fetuses have testosterone and other androgens. Most of that testosterone in both sexes is produced by the fetus, but maternal testosterone can cross the placenta and play a role. High levels of testosterone are usually due to male testes, but testosterone levels in both males and females can vary for many reasons, and it can vary from one day to another. Varied testosterone levels during different critical periods will lead to varied outcomes for each trait for both males and females. Just the same, more structures in males will become more masculine and more structures in females will become more feminine, given that male fetuses produce more testosterone over the long run.

6) While all that is going on, other elements in the microenvironment will also impact the outcome, especially during

Chapter 4—Apps

these critical periods. This includes environmental toxins and other hormones. This also includes many maternal factors such as antibodies, maternal hormones, and even maternal nutrition. Each of these factors might diminish or might augment the action of testosterone.

7) By the time the infants are born, each has a unique version of this app. Most of the male infants will be born with a more masculine version, and most females will be born with a more feminine version. However, given all of the factors at play, there will be a lot of variability. Some of the females will have masculine versions, and some of the males will have feminine versions.

8) However, it is not done yet. Next, the child's environment will continue to play a role. If the child grows up in a society without cars, then that child probably won't develop much interest in auto mechanics. If that child lives in a community where one can't function without understanding auto mechanics, then both males and females will likely develop some interest. If they are in a society like ours, where males are socially rewarded for knowing about auto mechanics, but females are not, then that will further reinforce the trend of increased interest among males. This particular app is a complex app and nurture/culture clearly play a big role.

Like most apps, it shares some brain structures with other apps. Earlier, I talked about the app for mental manipulation of 3D images. It seems obvious that this 3D app would also be relevant to auto mechanics. It makes intuitive sense that being able to mentally manipulate 3D objects might also make one more likely to be interested in things where that capacity is needed. So it seems likely that these two apps could be correlated, but the correlation won't be exact since there are also other factors that determine one's interest in auto mechanics.

Remember: Any particular man could be the most masculine guy in town and could still carry a feminine version of the auto mechanic app, thus having a divergent trait. Remember, we are all a mosaic of masculine and feminine traits.

In our example above, since more males are interested in auto mechanics compared to females, then it follows that a masculine version of that app will lead to more interest in auto mechanics and a feminine version will lead to less interest. We can assume that sex hormones contribute to this difference along with inheritance, prenatal environmental factors, and postnatal factors such as culture, social reinforcement, and education.

As I stated above, some will argue that some of these traits are only masculine or feminine because of what society dictates and that sexual differentiation of brains has nothing to do with it. I don't think that this is true of this particular example, but I accept that it can't be proven yet, and society/culture clearly plays a large role.

Keep in mind that social rules are frequently an amplification of real statistical differences that are hard-wired into our differentiated brains. If there is even a small difference between males and females on any trait, culture and society will reinforce and amplify it by giving social sanction to individuals for expressing the "correct" traits.

In the example above, if more little boys are innately interested in cars/planes/trucks than little girls, then little girls who do happen to be interested aren't encouraged to pursue that interest and are often actively discouraged. This ends up amplifying any statistical gap between the sexes. However, it does not eliminate the exceptions, and there are plenty of very competent female auto mechanics.

Example #6—innate interest in ballet dance

Let's reinforce how the process works with another example. This example is also intuitive and stereotypical. Take the app that determines how much a person will be interested in ballet. In this case, more females than males have a version of the app that makes them highly interested in ballet. A more masculine version leads to lower interest in ballet. Of course, there are plenty of very masculine men who have the feminine version of that particular app. We are all mosaics.

This app will follow the same process:

1) *Inherited DNA codes for the genes and gives a basic undifferentiated template for that app. Even before being acted on, the template may be intrinsically more feminine or masculine based on inheritance.*
2) *The regions of the brain that are involved with this app will each pass through critical periods.*
3) *During critical periods, the relevant tissues are particularly sensitive to testosterone levels, which will lead each tissue to masculinize or feminize, depending on testosterone levels.*
4) *Other environmental factors (such as other hormones, nutrition, toxins and antibodies) will also have an impact, especially during critical periods.*
5) *After birth, the individual's external environment will continue to act on this app, including the physical environment, as well as culture, learning, parenting etc.*
6) *After birth, the individual's internal environment will also play an ongoing role, especially their own hormones.*

In this particular trait, it is pretty obvious that culture, parental expectations, and education are going to play a huge role in the final outcome. This is going to greatly amplify any gender gap that has been hard-wired into the brain by the action of testosterone.

Masculine traits or feminine traits or in-between: a gradient

It turns out that for most apps that involve the brain, the usual outcome is a version that is neither extremely masculine nor feminine. During critical periods, testosterone levels (and other factors) will push it one way or the other, but they will rarely push it to the extreme.

Thus, each person ends up with a version that is on a continuum between masculine and feminine. One person's version might be highly feminine, and another's might be partly feminine and partly masculine. Their version could be 50/50 masculine/feminine, a 60/40 mix, a 99/1 mix, or any mix in between. So, in addition to the fact that we are all a mosaic of masculine and feminine traits, each of those traits falls somewhere along the gradient between masculine and feminine. This adds further variety to our mosaics.

Binary

Let's talk about "binary" and what that means. Binary implies that something is all or nothing. It is on or off. It is one thing or another. A light switch is binary, on or off—unless it has a dimmer switch. A dimmer switch is non-binary.

Most traits that are sexually differentiated (or gendered) are not fully binary. If it were fully binary then everybody would have either a fully masculine version of that trait or a fully feminine version.

Just to illustrate what I am talking about, I will take an example that is totally cultural: hair length.

If we had a town where every single male in that town has short hair and every single female has long hair, then you could say that hair length is fully binary in that town. Everybody has either the masculine version or the feminine version of that trait.

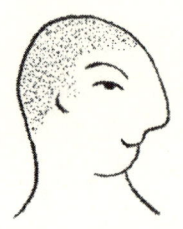

Now suppose that in another town, most of the women have long hair and most of the men have short hair, but some of the women in that town have very short hair, and some of the men have very long hair. In this scenario, the trait is still fully binary. Everybody has either long or short hair, so everybody has either the masculine version of that trait or the feminine version, even if the version they have doesn't always match the sex of the individual (divergent traits).

In yet another town which is more typical, some people have short or long hair, but some people have medium-length hair. This includes some males and some females. Since there are a

Chapter 4—Apps

lot of people in this town with a version of this trait that is in between masculine and feminine, then we would say that this trait is much less binary in this town. In this town, there is a gradient from masculine to feminine, and everybody's hair length is somewhere on the gradient.

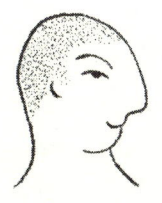

This is useful to think about when we talk about other masculine and feminine traits, especially the ones that are based in our brains and ask ourselves just how binary they are… or aren't.

There are no sexually differentiated traits that are 100% binary, but some traits are more binary than others.

Let's look at the most binary trait I can think of, which is our genotype. Our genes control most of our traits, but you could even say that our sex chromosome configuration is a trait in and of itself.

A masculine version of this trait is to have one X and one Y sex chromosome, and a feminine version of this trait is to have two X chromosomes. This is as binary as you can get, but even this trait isn't completely binary because you do have some people with different chromosome combinations. If someone has XYY (one X but two Y chromosomes), you could argue that this version is even more masculine and that XYYY is even more masculine than that. There are also individuals with XXY (two X and one Y chromosomes) and you might argue that they fall more in the middle of the spectrum. There are also individuals who are XX, but a piece of Y chromosome was attached to the X chromosome so one of their X chromosomes has all the important Y chromosome genes. That person could develop in a way that is hard to distinguish from your average XY male, so you would have to place them somewhere on the masculine side on the spectrum. Even though these cases are rare, they exist and help illustrate how it is hard to find a gendered trait that is fully binary. See Chapter 6 for more details about these chromosomal conditions.

For another example, let's look at the apparatus that determines the length of the penis/clitoris. This is interesting because the penis and clitoris both grow from the same structure in the fetus. A particular set of cells will grow and if it grows a lot, it becomes a penis and if only grows a little, it will be a clitoris. Sometimes it grows somewhere in between, and you can't determine whether it is a penis or a clitoris. This is a trait that is strongly binary but not fully binary. Most clitorises are substantially smaller than most penises. Meanwhile, there are exceptions and overlap. There are enlarged clitorises and very small penises. There are also people who are intersex. Some of them are born with genitalia that are between a penis and a clitoris.

Chapter 4—Apps

Traits that are less binary

Let's look at height. Height is a trait that is even less binary. If we assume that there is a developmental apparatus that controls height, a more masculine version of the app will produce a taller person and a more feminine version of the app will produce a shorter person. But with this trait, we have a lot of overlap. There are a lot of tall females, and some females are taller than most males, while some males are shorter than most females.

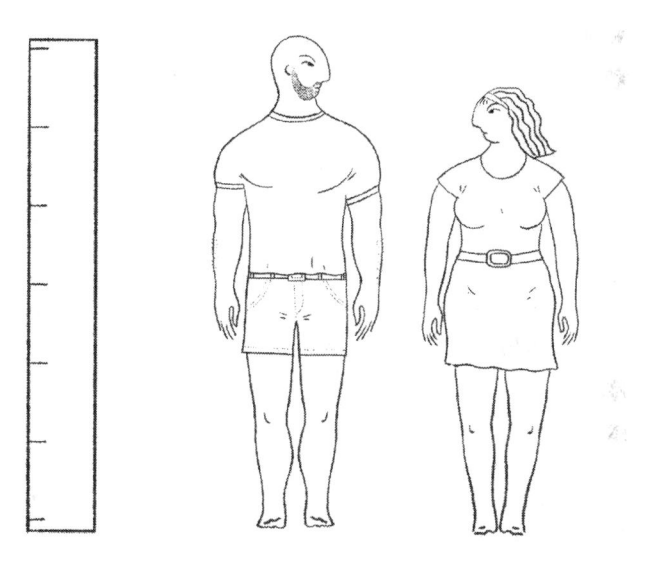

When it comes to height, there are a lot of factors that will impact your height, such as nutrition and genetics, and these factors actually play a bigger role than your sex. But your sex and your sex hormones do play a role. In the end, sexual differentiation does not lead to a fully binary outcome but does lead to a statistical difference between sexes.

The apps that control interest in auto mechanics or interest in ballet are also much less binary, and many people fall in the middle of the gradient between masculine and feminine with both of these traits. In fact, most traits based in our brains are less binary, meaning intermediate outcomes will be the norm. However, some brain-based traits are more binary than others (e.g., the capacity for violence, which we will discuss later).

The role of thresholds in highly binary traits

The mechanism for development of highly binary traits likely involves "thresholds".

For a highly binary trait, there will be a standard feminine development regardless of testosterone levels, up to a particular threshold. If testosterone levels surpass that threshold during a critical period, it will trigger a switch to a standard masculine development (for that trait). Higher testosterone beyond that level will have no further impact. In this case it is like a light switch, an all or nothing response. Since typical male fetal testosterone levels are many times higher than typical female levels, there is a predictable outcome for these binary traits.

Adding further complexity, it is still possible for some of these highly binary traits to have an intermediate outcome if that person's testosterone levels happen to be very close to the threshold (during critical periods). Even though most females will be substantially below the threshold and most males will be substantially above, there might be a small minority who will have testosterone levels hovering near the threshold, triggering an intermediate outcome in some cases.

Also, remember that each particular trait has many different genetic templates. For a particular trait, some families might carry a highly binary version, and others might carry a less binary version. The families that carry less binary templates for a particular trait will have more individuals with intermediate outcomes for that trait.

Furthermore, different versions of the genetic template might have higher or lower thresholds. If it has a lower threshold, it will masculinize more easily because it will require lower testosterone levels. If has a higher threshold, it will require higher testosterone levels to trigger masculinization.

Therefore, a particular trait might have a genetic template running in certain families where the threshold is lower than average and thus easier to reach. Females in those families are more likely to masculinize that trait. Since their testosterone levels can fluctuate, it is more likely that they will happen to have high enough levels during a critical period for that trait, triggering the masculinization of that trait. You could say they inherited a template that is more "sensitive" to testosterone levels.

Conversely, a trait might have genetic templates running in certain families where the threshold is higher than average and harder to reach. The threshold might be so high that some of the males won't masculinize that trait. In these males, testosterone fluctuations during the critical period for that trait lead to levels that fall below the threshold. Thus, masculinization won't be triggered, and they will develop a feminine version of that trait. You could say they inherited a template that is more "resistant" to the effects of testosterone.

In other words, having a higher threshold (for triggering masculinization or a trait) is a type of testosterone resistance, and having a lower threshold is a type of testosterone sensitivity. Sexual orientation is a good example of a trait that likely uses this kind of mechanism. This would help explain why homosexuality can run in families, as we will discuss in Chapter 5.

For those who enjoy statistics and graphs, any completely non-binary trait will likely have a standard bell curve with the peak in the middle. A highly binary trait will have two separated peaks with relatively few individuals falling in the center.

Example #7 and how it develops—baseball throwing

Chapter 4—Apps

Now let's show some more examples of how these "human apps" develop. Let's look at an app for throwing a baseball.

There is a masculinized version of this app. Those with this version are very effective at throwing baseballs. They can throw them farther, faster, and more accurately. Most men have a masculinized version of this app, but so do plenty of women. So how does it all work?

This app works as an interaction between several different regions of the brain and the body. The brain creates, modifies, and sends the signals that coordinate muscle activity. The muscles execute the action. The eyes are essential in determining the aim. The vestibular apparatus in your ear is essential in balance and body position. The proprioceptors located throughout your body help determine your arm and body positions at any point in time during the throw.

One factor for throwing is the differentiation of the bones and the muscles. As each fetus grows, they will develop basic bones and muscles in their arms. In childhood, these bones and muscles won't have size differences due to gender—that won't come until adolescence. Then, there is a critical period for sexual differentiation of bones and muscles that happens during and after puberty. During this critical period, if there are high levels of testosterone present, as would be the case with most males, you will have substantially more growth of the muscles and bones.

Thus, at puberty, masculinization will lead to longer arm bones, with bigger muscles, and attach them to a bigger overall frame. This will help lead to a stronger throw. Lots of other factors will impact how big the muscles and bones will grow, including inherited genes. However, any individual who is exposed to high levels of testosterone during this critical period will develop bigger muscles and bones compared to their sisters who likely won't have as much testosterone.

You will see this pattern a lot. High testosterone is going to masculinize that trait. In this case, the presence of large amounts of testosterone from ages 12–22 will result in bone and muscle growth exceeding what would happen in an individual without that exposure. This is why, on average, men's arms are bigger, longer and thicker compared to women's.

But that is not the whole story. Boys, on average, are already outperforming girls with baseballs well before puberty. It is not just their bigger muscles and longer bones. There is also a brain apparatus that contributes to this capacity.

The way that the arm and core body move together is crucial to throwing a projectile. These movements are coordinated in the motor areas of the brain. There is a "human app" in the brain that controls ball throwing, and coordinates the muscles involved.

Everybody knows the stereotype of what it is to "throw like a girl." Everybody also knows that not all girls "throw like a girl" and that some boys do "throw like a girl." Most people mistakenly believe that a boy might throw "like a girl" because they weren't taught how to throw. But in reality it is mostly due to the apparatus in their brain that makes up their "human app" for throwing objects. In the case of ball throwing, boys tend to throw objects coordinating arm extension with torso twisting and stepping forward to give substantially more force to the throw. Girls tend to throw with just the elbow as you would in dart throwing, without using the torso, which ends up being much weaker. The girls (and the boys) who don't use their torsos can be taught to do so, but it doesn't come as naturally for them and requires more practice.

Various apparatuses in our brains coordinate our muscle movements in ways that result in visibly different ways of moving that could be described as more masculine or more feminine. It turns out that each way of moving might have different advantages for different activities. It makes sense that a movement strategy that makes a person effective at throwing baseballs might not be so good at helping them dance gracefully. In fact, it might make them much worse at dancing gracefully.

Thus, a masculinized app that helps someone coordinate the optimal motions for throwing might be a detriment for other types of movement, such as ballet dancing.

Remember, this is just an illustration. There might be people out there who are quite good at both ballet and throwing baseballs. I am just trying to pick examples that are intuitive to illustrate a general principle. There are lots of factors that determine how good you are at throwing a ball or dancing ballet. Some of them are learned and some of them depend on family traits that could impact both sexes. However, the sexual differentiation caused by testosterone is certainly an important factor.

This brain region or app that controls how we throw a baseball develops following the usual pattern, and we will keep repeating this story. A template develops in the brain of every fetus based on their genes, inherited from both parents. This template will eventually develop into the app that determines how they throw. Even before sexual differentiation occurs it will already have some of its characteristics. Then, during a critical period, which is a set period of time early in its development, the presence or absence of high levels of testosterone will determine whether that structure will masculinize or not. In this case the critical period is probably during mid-pregnancy when most brain structures are developing and differentiating.

The difference in ball throwing and other movements is already evident in childhood well before puberty. We know that during most of infancy and childhood, there is not much testosterone activity. This helps confirm that the critical period for this app occurred in utero. For male fetuses, testosterone levels are high in the prenatal period and briefly just after birth. After that, there is minimal testosterone secretion during childhood until puberty starts.

Chapter 4—Apps

Remember that some critical periods are very short. A critical period might be just a few days or even less. Nobody knows exactly when most of these critical periods are. We do know that most brain development happens layer by layer during pregnancy and that any sexual differentiation of brain tissues happens after the gonads have already formed by week 8.

For the sake of illustration, let's make a speculation. Let's speculate that the brain structure for ball throwing (and other movements) has a critical period at week 18 of gestation and lasts 48 hours. During these 48 hours, that structure would develop in a very masculine way, a sort of masculine way, a sort of feminine way, or a very feminine way. It would land somewhere along this gradient, depending on whether or not there are high levels of testosterone and on how sensitive/resistant that trait is to the effects of testosterone. Once that critical period ends, it will keep going down the path that was set during the critical period. This is a very common phenomenon in embryology and development.

So let's expound more on this process, along with some caveats.

You have a basic template that develops initially with the potential to masculinize or feminize. Before you even start, it might be more masculine or more feminine based on its genetic makeup, regardless of the sex of the individual, even before sex hormones start to influence the progress. But no matter where it starts out, testosterone will push it in a masculine direction if testosterone levels are high enough during critical periods.

However, it might be genetically more sensitive to testosterone, or it might be more resistant. If it is more sensitive, then testosterone will have a bigger impact on the final outcome. If it is more resistant to testosterone, then testosterone won't make as much difference.

So, if that person inherits a testosterone-resistant version of that app, then the inherited genes will play the biggest role, and there won't be much sexual differentiation. If a brother and sister were to inherit the same testosterone-resistant version of that app, they would have a very similar outcome. On the other hand, if they both inherit a testosterone-sensitive version of that app, then testosterone will lead to a larger gender difference and the brother will likely outperform his sister on ball throwing.

Testosterone levels during a critical period can be impacted by a number of factors. The biggest factor is whether or not the fetus has testes that are producing testosterone, as would be the case for most male fetuses. However, there can be other factors. They might be living in a uterus with a mother who is producing her own testosterone or other analogous hormones that are leaking through the placenta enough to play a role. Or they might be living in a uterus inside a mother who is impeding the action of testosterone by the chemicals that she is secreting across the placenta. Those chemicals can include environmental microplastics and toxins that might inhibit the effect of testosterone (or that might do the opposite and exaggerate it). They might be a female fetus with a condition where they are producing excess androgens that mimic

Chapter 4—Apps

testosterone, even though they don't have testes. There might be other factors that are slowing down or speeding up the testosterone production in that particular fetus on that particular day.

So yes, the biggest effect comes from the fetus itself, but there is an important environmental effect, meaning the environment provided inside the mother's uterus. This is impacted by the mother's diet, exposure to toxins, immunologic response to pregnancy, allergies, stress, etc. Since the path is set in an irrevocable way during the critical period, sometimes the maternal and environmental factors are determinative.

Later, this baby is born with a brain structure that is going to dictate how capable they will be at throwing a baseball, even though they won't touch a baseball for some time yet. In the next few years, they will start practicing throwing things. If they are less interested in throwing, then they aren't going to get quite as good at it. If they are encouraged to throw or play those kinds of games, they will likely improve regardless of the version of the app they carry. If they get a neurologic illness or a serious injury, they may lose that advantage. But overall, it won't take long before their innate capacities will be evident. For most males, their muscular growth at puberty will subsequently amplify their throwing advantage.

In summary, the brain structures that coordinate how we move have sexual differentiation. Some people move like ideal rugby players, but those people will likely not be the best ballet dancers. Among athletes, the best tennis players aren't the best basketball players. Neither of those will be the best figure skaters.

Androgens like testosterone may or may not give an advantage for any particular capacity in sports, but when it comes to muscle strength, androgens universally give an advantage. This is why sporting bodies forbid "doping" which is a way of artificially masculinizing the body so that it can have more strength and provide a competitive edge in sports that require more strength.

Example #8—empathy app

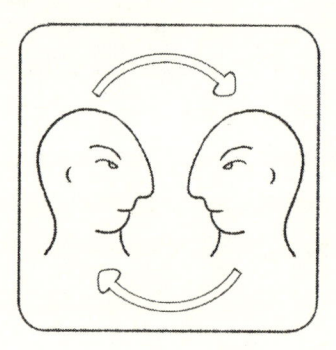

There is a brain apparatus for empathy. Like other apps, it is located in diffuse areas of the brain and includes some specific brain structures and interconnections.

It turns out that this app is also gendered. On average, females show a higher capacity for empathy and perform better on empathy tests—for example, the ability to identify the emotion

Chapter 4—Apps

of an individual based on a photo of their face. Females tend to outperform males on this skill.

This gender gap is referred to as the "empathy gap". This app is almost certainly related to the above-mentioned face-gazing app that determines the interest a particular infant will have in looking at the faces of their caregivers. It makes sense that spending more time looking at faces would eventually make you better at reading those faces. This is another example where elements of one app can contribute to the development of another.

Just like the ball-throwing app, this empathy app starts with a template that is inherited and can be inherently more masculine or feminine regardless of the sex of the individual. However, this app will pass through one or more critical periods and, depending on its sensitivity to testosterone, it will masculinize or feminize based on the levels of testosterone present on those days. Later, life experiences will have an ongoing influence on that app.

Like most brain traits, most people will end up with a version of this app that is in between the extreme of masculine/feminine. On average, more females will have a feminine version and more males will have a masculine version. However, there will be plenty of males with feminine versions and plenty of females with masculine versions. Like the other apps, there are a lot of exceptions and you cannot predict the sex of any individual based on how they perform on empathy tasks.

This example can also help illustrates another important trend. Even though an extremely masculine version of this app would mean having little or no capacity for empathy, very few males actually have this extreme version. In fact this is considered pathological. In reality, the vast majority of both men and women have a high capacity for empathy. The difference is in the average. You might say that the average male has a high level of empathy, and the average female has an even higher level. But there is so much variation and overlap that it is impossible to predict based on an individual's sex. A lot of the other apps we have discussed here work this way as well.

Example #9—capacity for violence app

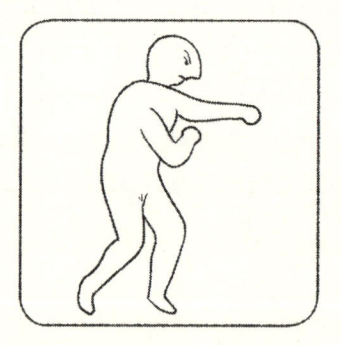

Capacity for violence is another gendered brain/personality trait. We will discuss it here because it illustrates several important concepts.

This app happens to be a fairly binary. There is overwhelming evidence that males are more violent than females. Thus, a masculine version of this app indicates a higher tendency toward violence. Like all other apps, this app is not monolithic. In fact, most men are not violent, but among those who are violent, the vast majority are males.

Like all the apps described, the development of this app, and its subsequent expression is a product of both nature and nurture. Most of us want to minimize violence in our society, so we really need to understand all of the factors that contribute to its expression. This can help us explore possible interventions that could help us reduce violence.

As a starting point, each fetus will inherit a template that is coded in their genes, and this template is the first factor that will contribute to the final outcome. That template will code for brain tissues involved with this app. These brain tissues will subsequently have critical periods where they will be pushed in a masculine direction or feminine direction based on the presence or absence of high testosterone levels during that critical period. As usual, environmental factors will play an outsized role while this is happening.

We probably agree that it would be unreasonable to interfere with the normal male development in order to prevent future violence. However, a substantial amount of future violence is a result of abnormal development. This gives us an excellent opportunity to intervene. Basically, the fetal environment needs to be protected. Exposing a fetus to toxins or poor maternal nutrition can lead to developmental issues that predispose that individual to future violence, regardless of their sex. A fetus gestating in a mother who is experiencing violence or extreme stress will also lead to a higher risk of future violence. These risk factors are highly associated with poverty and pollution. Thus, we have a chance to decrease the expression of the "violence app" by simply protecting the fetus from the effects of pollution, poverty, and violence.

After birth, the environment continues to play an important role in the development of this app. In fact, childhood experiences have a particularly strong impact on future violent behaviors.

Toddlers of both sexes will show aggressive behaviors, although boys are already showing higher physical aggression on average compared to girls. However, with time, children's brains gradually develop better capacities to inhibit these impulses. As they get older, they gain better control of their violence. This capacity to inhibit violence could be considered a separate app that impacts the "violence app". I will call this the "violence-inhibiting app".

This "violence-inhibiting app" is vulnerable. Insults to brain development related to toxins, injury or malnutrition can easily impair this important ability to inhibit violence in both males and females. Meanwhile, childhood trauma, especially witnessing or experiencing violence, greatly increases the risk of future violence. Instability and neglect also increase the risk.

Environmental interventions, both prenatally and during early childhood, offer excellent opportunities to decrease future violence. Our best chance of decreasing the potential for future

Chapter 4—Apps

violence is simply protecting the fetus AND the child from the damaging effects of pollution, poverty, and violence.

Puberty is the next big fork in the road. Some individuals will arrive at puberty with a violence app that is primed for violence by their genetics, their development, and their environments. Those who also have functioning testes will start to produce high levels of testosterone. Their brains will be bathed in testosterone, including the brain tissues involved in the violence capacity app. This testosterone exposure will induce further masculinization that will increase their capacity for violence. Due to the effects of this testosterone surge, violence typically peaks in affected individuals during late adolescence and early adulthood.

In spite of this pattern, most males will never engage in substantial violence. They will exit puberty with a capacity for violence that is well balanced by their capacity to inhibit that violence. However, a small percentage of these males will have a violence app that predisposes them to high levels of violence and an impaired violence-inhibiting app. These individuals will be responsible for most of the violence in society.

Fortunately, for most of these males with a high capacity for violence, the risk of violence recedes with time. As they mature, rates of violence will fall as brain development proceeds, allowing fuller development of the violence-inhibiting app. The testosterone surges of adolescence and early adulthood subside, which also leads to a decrease in violence. This is a demonstration of an app that was highly influenced by testosterone. Some of that influence is permanent, while some of the effect diminishes once the testosterone is withdrawn or decreased.

Female violence will be much less common, corresponding to the fact that very few females will have this level of testosterone exposure. Just the same, there will always be examples of female violence. In females, most violent tendencies will be due to developmental insults that impede the development of their brain's "violence-inhibition app." In a smaller number of females, masculinization of their "violence app" due to testosterone effects during gestation could also play a role. Importantly, females won't get that puberty testosterone surge. This is probably the biggest factor that results in dramatically lower rates of physical violence among females.

Example #10—sexual orientation app

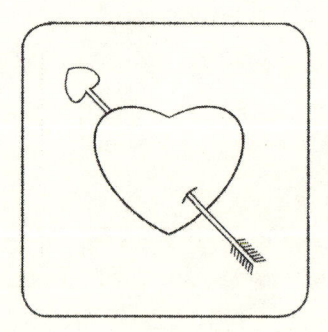

We are going to look at two other traits that are less binary than chromosomes and less binary than genitalia but are of great interest to a lot of people.

The first of these is an apparatus in our brain that determines our sexual orientation. For sexual orientation, a feminized version is an orientation toward males. Most, but not all, females have a version that is feminized and are therefore oriented towards males and are heterosexual. However, plenty of them have a version that is masculinized, and they are more oriented toward females and are therefore homosexual. You also see that a lot of people fall between these extremes and land in the bisexual or pansexual realm.

Most males have a masculinized version of this app, giving them a sexual orientation toward females. Meanwhile, there are plenty who have a feminized version and are oriented toward males, meaning they are homosexual. There are plenty who are in between and might be characterized as bisexual or pansexual.

Example #11—gender identity app

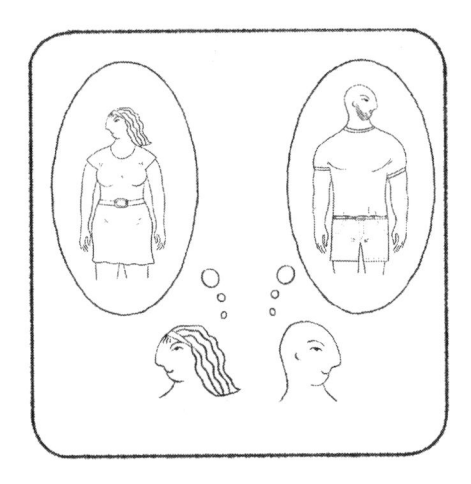

Another separate but similar app is our gender identity app. We have an apparatus in our brain that determines which gender we see ourselves belonging to. If this app is masculinized then that person will identify themselves as a male. If it is feminized that person will identify themselves as a female.

However, like most traits, this app is not fully binary. Most people see themselves as male or female, and fewer people see themselves as fitting somewhere in between, although those people clearly exist in substantial numbers. Like the app for sexual orientation, some males end up with the feminine version and some females end up with the masculine version.

Looking again at definitions

Let's focus on that last sentence for a minute.
<u>Some males end up with the feminine version, and some females end up with the masculine version.</u>

This is a place where the definition of male and female needs to be clarified, as our assumed version might lead to confusion.

When I was talking about all the other traits, it didn't matter which definition of male and female I used because it didn't impact the veracity of the statement.

When it comes to describing the gender identity app, then I am more likely to need to clarify.

Let me make a highly precise version of that sentence.

The original sentence:
<u>Some males end up with the feminine version, and some females end up with the masculine version.</u>

A highly precise version of that sentence:
<u>Most people who are born with a penis (or what appears to be one) also get the masculine version of the gender identity app, but some people who are born with a penis will get a feminine version of the gender identity app. Most people who are born without a penis also get a feminine version of the gender identity app, but some people who are born without a penis will get a masculine version of the gender identity app.</u>

Categories of gender identity

As you see, some people have a divergency on this trait. These people are called "trans" because of this divergency. People who don't have this divergency are called "cis." These are scientific prefixes used to indicate "being on the same side (cis)" versus "being on opposite sides (trans)," so these prefixes have been adopted to describe what I am calling a feminine/masculine divergency (similar to how I describe other traits as having frequent divergencies).

Thus, we end up with the following categories:

Trans people:
1) Assigned male at birth, usually based on the presence of a penis, but having a gender identity app that is fem-

inine, meaning that they have an internal/innate sense of being female, or
2) Assigned female at birth, usually based on the absence of a penis, but having a gender identity app that is masculine, meaning that they have in internal/innate sense of being male.

Cis people:
1) Assigned male at birth and having a matching (non-divergent) gender identity app that is masculine, or
2) Assigned female at birth and having a matching (non-divergent) gender identity app that is feminine.

Non-binary people:
1) People who are assigned male or female at birth but their gender identity app is neither fully feminine or fully masculine, so they don't identify as being solely male or solely female.

We will discuss sexual orientation and gender identity more extensively in the next chapter.

Chapter 5—Orientation and Identity

Parameters of sexual orientation

The big question: why are some people gay when most people are straight?

How does that develop?

Remember that we are all mosaics of masculine and feminine traits. Part of this mishmash is coded by our genes. A lot of our traits are just inherently more masculine or feminine even before they differentiate, simply due to inheritance. Furthermore, some of these genes will carry versions of the trait that are more resistant (or sensitive) to the effects of testosterone compared to the versions other people carry. On top of that, due to the different timing of critical periods, some of these structures will get higher or lower testosterone exposure due to random fluctuations or other factors that are present in the fetal environment on those particular days. All of those contribute to a huge amount of variation. It doesn't change the fact that most men will have more masculine traits due to their high testosterone production. However, even fully functional testes will not lead to a fully masculine outcome in every trait. It also doesn't change the fact that most females will have more feminine structures due to their lower testosterone levels. However, even without testes, each female will end up with some brain structures that are more masculine.

All of this applies to our sexual orientation app.

Remember, we each have an app that controls our sexual orientation. Like all apps, it is made up of diverse regions of the brain that act together to control our sexual attractions and sexual aversions. Let's review how that works.

When there is sexual attraction, a person might suddenly seem interesting and spark that sexual interest. However, before we get to that point, our brain has done a lot of legwork. First, our eyes, ears and nose take in a lot of information about each person that we encounter. Our brain will discern a lot of data about them. There will be ongoing assessments of sexual attraction vs. sexual aversion. Our brains will address many relevant questions about that individual, such as

1) whether they are the correct species,
2) whether they are too close of a relative,
3) whether they are in a good state of health,
4) whether they have physical characteristics that we are attracted to,
5) whether they have personality characteristics that we are attracted to,
6) whether they are in the correct age range,
7) which gender,
8) etc.

From some of these basic questions, most the people are already ruled out by at least one of these criteria, and without any conscious awareness, our sexual attraction apparatus will go back on standby. If, for some reason, we are compelled to consider a person that doesn't attract us, we might become conscious of our sexual aversion to that person. Thankfully, we don't have to consciously consider sexual attraction or aversion at every encounter because our brain filters out most of that for us unconsciously.

There are a lot of factors that contribute to sexual attraction or sexual aversion. Sexual orientation is not simply about the gender or sex of the other person. We are oriented towards a number of their traits. Just the same, gender does play an enormous role for most people.

Sexual orientation app

Let me remind you that when I say app, I am referring to a brain "apparatus" (which is similar to an application on our smartphones). Like any human app, this app includes the involved, diffuse regions of the brain and their interconnections. This app takes in information from our sensory organs (eyes, ears, nose, skin, etc.), processes it, and then sends output to other parts of our brains or to our bodies to impact our behaviors.

In the case of our sexual orientation app, we take in information about the person with our eyes, ears, noses, etc. The app uses relevant brain structures to process this information and determines our level of sexual interest in that person. It then sends output to other parts of the brain and to our bodies that determine our behavior in that situation. Depending on the attraction criteria that are built into our sexual orientation app, the output might motivate us to flirt with or pursue that person. It might lead us to feel strong emotions such as infatuation or love. It might induce sexual arousal. On the other hand, if we have aversion instead of attraction, it might lead to the opposite response.

A feminine version of the sexual orientation app will produce a person who is oriented toward males (androphilia). Thus, the object of attraction of somebody with a feminized sexual orientation app will generally be male.

A masculine version of this app will produce a person who is oriented toward females (gynephilia). Thus, the object of attraction of a person with a masculinized sexual orientation app will generally be female.

Most males end up with more masculinized versions of this app and are more oriented toward females. Most females end up with a feminized version of this app and are oriented toward males. Some males end up with a feminized version of this app and are oriented toward males. We refer to these males as gay males. Some females end up with a masculinized version of this app and are oriented toward females. We refer to these females as gay females or lesbians.

Chapter 5—Orientation and Identity

This particular app is not fully binary. Everybody falls on a gradient somewhere between extremely masculine and extremely feminine.

Those who fall near the middle have equal attraction to males and females. These people might fit the description of bisexual or possibly pansexual.

This starts to explain why there are straight people, gay people, and bisexual people. However, these categories don't fully explain the gradient. Even people who are predominantly straight might have some level of attraction to the same sex that puts them somewhere on the continuum between straight and bisexual. The parallel could be true of any gay person.

Also, these distinctions don't always take into account levels of aversion. A person could have a preference toward one sex but not have a particular aversion to the other. So even if they are not necessarily bisexual, they still might be capable of enjoying sex with someone of their non-preferred gender. Others might have a complete aversion to the idea of sex with someone of the "incorrect" gender.

When it comes to people who are gay, you could say that they are divergent with this trait (sexual orientation). Gay men are males, and this particular trait is not typical of males, so by the definition presented earlier, they have a divergency. We all have different divergent traits, so I am not calling these abnormalities. I am just putting them into a useful category for the sake of this discussion. Gay women are divergent in this trait because they are females who have a masculine version of this particular app.

The development of sexual orientation

So, let's review the process of how these traits arise in our developmental pathways. You are going to recognize the pattern.

If the fetus is male, at weeks 6–8 of gestation, its SRY gene will turn on the testosterone factory and the fetus will start producing androgens and develop testes. If the fetus is female their gonads will develop into ovaries. The fetuses with testes will start secreting testosterone, and their genitalia will develop in a masculine direction, and they will grow a penis. The female fetuses will lack the large testosterone surges and will have a different balance of hormones that will lead to the development of a vagina and clitoris. These processes have critical stages. For gonad development, it will be weeks 6–8 and for genital develop-ment it will be weeks 8–12. By the end of week 12, both gonads and genitals are permanently on the path that they will stay on. That fork in the road has been passed. No amount of testosterone can change the path after the critical period ends. Testosterone levels can still have an influence and can impact further growth of the penis. However, the penis, vagina, and clitoris all have their architectures established by week 12. Meanwhile, different areas of the brain have much later critical periods. Our brains continue to develop and differentiate after week 12 and throughout the pregnancy/gestation. Some of that development even continues after we are born and even into adulthood. Thus, the critical

Chapter 5—Orientation and Identity

periods of sexual differentiation for the different parts of our brains happen later than they do with our gonads and genitalia.

The sexual orientation app is one of these areas. Its critical period could be weeks later than the critical periods for gonads and genitalia. That is part of the reason there can be divergencies. The conditions at play can be very different on those critical days that will determine the direction of their sexual orientation compared to how they were during the critical days of gonad and genital differentiation.

Male fetuses generally produce large amounts of testosterone starting at week 8. Their testes are already formed and are doing their job. The high levels of testosterone induce masculinization of any brain apps that are subject to sexual differentiation. Since the levels of testosterone can vary from day to day for a number of reasons, then the amount of masculinization can vary based on the level of testosterone present during any critical period. For most males, the sexual orientation app will differentiate in a climate of high testosterone, so most males will end up with a masculine version of this app and be oriented toward females. For most females, their sexual orientation app will differentiate in a climate with low levels of testosterone, so they will end up with a feminine version of this app and be oriented toward males.

For gay males and gay females, a masculine/feminine divergency occurs with their sexual orientation app. Several different mechanisms can be involved in this divergency. These mainly involve the interplay between inherited genetic factors and testosterone action during critical periods.

Some males might inherit genes from their parents that leave them more likely to develop a feminine version of that trait. There are different gene combinations that could increase the likelihood of that outcome, and each combination might have a distinct mechanism. One highly likely mechanism would be a set of genes that code for a sexual orientation app template that is more resistant to the activating effects of testosterone. Those fetuses would need higher levels of testosterone in order to induce a masculine outcome. Male fetuses with this gene combination are more likely to have days where their testosterone isn't high enough to induce masculinization. If the structures making up the sexual orientation app have a critical period during one of those days of insufficient testosterone, it will have a more feminine outcome, and that male will end up with a homosexual orientation.

There are random factors that can impact the level of testosterone. Variation in the amount of testosterone that their testes secrete might result in lower testosterone that particular day. Factors from the mother can play a role on that particular day, including antibodies, toxins, maternal hormones, etc.

For females, there are also combinations of genes that will impact the extent of feminization of their orientation app. There are different mechanisms involved in different individuals. One highly likely mechanism would be to have a genetic template for that app that is more sensitive to testosterone and will masculinize more readily. The levels of testosterone needed to masculinize the app would be lower. In this case, there would be a bigger chance that they will have days where their testosterone levels are sufficient to induce masculinization. If a critical period for their sexual orientation app arrives on one of those days that their levels happen to be high enough, then it will have a masculine outcome, and that female will end up with a homosexual orientation. Different factors could cause increased testosterone (or other androgens) to be present on those days. Those factors might be hormones they produce themselves or that the mother contributes. Other hormones, toxins, antibodies and other maternal factors can also impact the outcome.

With sexual orientation, there will be a lot of variability in the outcome. Most everybody will fall on the gradient between a fully masculine version and a fully feminine version. It isn't just the gay people who will have this variability. Straight people will also inherit variability in the baseline masculinity vs. femininity of their template for that app. They will also inherit diverse sensitivities to the effects of testosterone and other hormones on their app. They will also have fluctuations in testosterone levels and other factors during the critical period for that app. These factors will result in a sexual orientation app on a gradient between fully masculine and fully feminine, even among people who are straight.

There is also a continuum of sexual aversions, and everybody will end up at a different point on the gradient. All people, both gay and straight, will have varying degrees of aversion to sex with their non-preferred gender.

Another way that the continuum of our sexual orientation app is manifested is the large variability in our levels of attraction to different feminine or masculine features, even if we are primarily gay or straight. For example, a female might be straight, but might be more attracted to feminine men or certain feminine traits in a man. Some females might even have an aversion to highly masculine features such as a hairy chest or exaggerated musculature. A different female might be highly attracted to these features. Any individual will end up with an attraction to some masculine features and some feminine features, even if they are mostly straight (or gay). These tastes surely arise in the same way as our orientation, involving an interaction between our inherited genes and testosterone levels during critical periods of brain development, but these tastes will also be impacted substantially by cultural influences and other ongoing environmental factors.

So, when the baby is born, they will have a brain with an orientation apparatus that is mostly fixed. However, this sexual orientation app won't be fully active yet, and sexual attraction will remain relatively dormant until puberty. Just the same, our infant brains are primed to take in information and start to categorize things. One of the things we start categorizing is the genders of other humans. We are primed to be interested in the gender of each person we meet and try to fit them into a category. Our brains will draw on this categorization when sexual attraction emerges as a force later in life.

Chapter 5—Orientation and Identity

Sexual orientation and genes

There are different mechanisms that can play a role when it comes to developing a homosexual orientation (a divergent trait).

Remember, anytime I say divergency, I am referring to a feminine trait in a male or a masculine trait in a female. These traits are "mismatches" in that sense, but it is completely normal to have divergent traits, and every human has many, many divergencies.

In the case of the divergent (or homosexual) sexual orientation, there are different developmental pathways that lead to this outcome. There is not just a single cause for homosexuality. There are a wide variety of causes. You can't even say there is a single cause in any one particular person. What you have is a set of causes. It is also an interplay between the genes and the prenatal environment.

The first factor is the genetic makeup. There is not a monolithic gay gene. However, you can inherit a combination of genes that might increase your chances of having a homosexual orientation. In other words, you can inherit a "potential" for a homosexual outcome. Different genes have been found on different chromosomes that are associated with a higher chance of homosexuality. However, there is no single gene that is present in all homosexual males, nor is there any single gene that is present in all homosexual females. Furthermore, even those genes that have an association with homosexuality are also present in heterosexual people, so it is clear that there is no gene that will guarantee a homosexual outcome.

Identical twins teach us a lot about homosexuality and genes. Since identical twins have the same genetic makeup, we can get a better idea how big of a role genes play.

It turns out that if one twin is gay, then there is a higher chance that the other will also be gay. Different studies show different levels of agreement on this, but studies show that if one of the identical twins is gay then there is a 20-50 percent chance that the other will also be gay.

What this demonstrates is that both twins inherit a genetic possibility or "potential" to be gay. This genetic potential would be a set of genes that would give them a statistically significant chance of turning out gay. The difference in outcome is subsequently explained by a divergence in their environments. These twins are encountering somewhat different conditions during prenatal development. These differences have outsized importance during critical periods for the sexual orientation apparatus.

It seems counterintuitive that the identical twins would have a different prenatal environment since it seems that they share the same intrauterine environment. However, highly localized factors play a role in their developing brains. Each fetus is attached to the uterus in a different position. This leads to different access to nutrition from the mother. This is why twins often have different birth weights. They also have different access to other maternal factors, such as antibodies, maternal hormones, or toxins. These can all contribute to the difference in outcome. It is also possible that the timing for critical periods gets offset. The different nutrition and growth rates might mean that one fetus reaches certain stages at a slightly later date. This could lead to one fetus having a critical period on a different day than their twin, and the environment might be different on that day. This could mean that random fluctuations in testosterone levels could lead to one of the twins being exposed to lower levels during the day(s) when their sexual orientation app is going through a critical period, causing it to go down a different fork in the road (compared to their twin).

The different outcomes in many identical twins really demonstrate the fine balance that comes into play when the regions of human brains are sexually differentiating during critical periods.

As I pointed out earlier, everybody has every gene needed to have a feminine or masculine outcome for any trait. That means that, technically, anybody has a genetic potential to turn out gay or straight.

Our sexual orientation is determined by the interplay of a number of genes, so the particular set of genes somebody inherits will impact whether or not they have a significant potential to turn out gay. In the end, most people have only a small potential for a gay outcome. The majority simply have a set of sexual orientation genes with a more binary response to testosterone. They simply go one way if there is a lot of testosterone and the other if there is not, so they are highly likely to turn out straight. On the other hand, those with a significant potential to turn out gay have a set of genes that lead to a more variable response to testosterone levels.

For example, if they are a male fetus, but their developing sexual orientation app is resistant to the effects of testosterone, then it will only masculinize if there are higher than average levels of testosterone during the critical period. So, some of these males with this set of genes will turn out straight and some will turn out gay. This is one possible mechanism to explain why identical twins might turn out differently. If their critical periods are offset by a few days, and their testosterone levels are fluctuating, then they could end up with different responses and different outcomes.

In the case of females, they might end up with a version of that apparatus that is highly sensitive to testosterone. It might be so sensitive that even the levels of testosterone present in a female fetus might be enough to masculinize that apparatus. Since there is some variability in testosterone levels, some of them with that set of genes might turn out gay if there is enough testosterone during the critical period, while others will turn out straight if testosterone levels are lower on those days.

An interesting phenomenon in gay males is the big brother effect. What this means is that for a certain combination of genes, there is a birth-order effect impacting the chances of turning out gay, and it is based on the number of older

Chapter 5—Orientation and Identity

brothers. Basically, the more older brothers the male fetus has, the more likely that he will turn out gay.

This is a nice illustration of prenatal environmental effects—also referred to as epigenetic effects. It is not fully understood exactly how these effects work, but somehow the mother's body has an imprinted memory of how many male fetuses she has carried, and this memory is used to impact her current pregnancy. It has been theorized that it is an immune reaction against some male factors (involved in masculinization), and this immune reaction gets stronger with every male child she carries.

There is a separate pathway toward male homosexuality that appears to use yet a different combination of genes. This one appears to be related to left-handedness, which is more common in gay men than straight men. A curious observation is that gay males who are left-handed don't have the above-described birth-order affect. This makes it clear that there are different pathways that lead males to a homosexual outcome. This particular pathway involves a set of genes that interact with the part of the brain that determines handedness and also the part of the brain that determines sexual orientation but is not susceptible to birth order. Similarly, the birth-order pathway is not related to left-handedness (although somebody could have both pathways). There are certainly other pathways as well. Each pathway depends on a distinct combination of genes that give that potential, and then factors that are present in the prenatal environment will make the final determination.

Gender identity

Our gender identity app develops in a similar way. Our gender identity app is an apparatus spread out in different regions of the brain that work together to give us an innate sense of whether we are male or female.

We aren't alone in the animal kingdom to have this. In many species, we can observe gendered behaviors. The animal may not be consciously aware of their gender in the same way we are, but their brain is definitely tuned in to which set of gendered behaviors to express, whether it be male or female. For example, male bighorn sheep group into male herds. Each male will have an instinct to herd with other males and have instincts to engage in male behaviors. The females will have brains that inform them to engage in female behaviors and herd with other females. They don't have mirrors that help them figure it out. They know it intrinsically.

Thus, animals have their own version of this gender identity app. At some level, they recognize their own gender and they fit nicely into their gender roles. Another example is cats. In their own way, male cats recognize that they are competing with other males. Female cats know that they can mostly cooperate with other females but should mistrust males. The male cats don't have to look at their penises to figure out they are males. Male cats just know they are males (in their way), and they follow their urges to act like males. They have a structure in their brain responsible for making that happen. We can't know how conscious cats are of their own sex, but we do know that cats are acutely aware of the sex of every other cat they encounter. It is clear that they have a brain structure that guides them toward their own gendered behaviors as well as their gender-based interactions with other cats.

In the case of humans, our gender identity app is much more conscious. Compared to cats, humans have a much more complex set of behaviors, and masculine/feminine behaviors are much more culturally influenced. We have a more conscious preoccupation with our gender identity, and this ends up guiding our behaviors.

Very early, each child will begin to categorize those around them as male or female and also will place themselves into one of those categories. They will then actively pursue activities and interests that they perceive as belonging to that gender.

Toy manufacturers have figured this out. There is nothing about the color pink that is gendered, but for the past 100 years pink has been associated with girls in Western society. At a very young age, our children figure out that pink is a "girl's" color. This is purely cultural. However, the desire to conform with gender expectations is not cultural—it is more inborn, although it varies from person to person. This leads to most girls being attracted to pink toys simply because they recognize that that is a culturally sanctioned "girl thing" and most boys will be averse to pink toys for the same reason. This tendency to favor activities that they observe as being "correct" for their gender is hard-wired into them, and it involves their gender identity app. Obviously, there is a wide variety in how strongly this tendency is present in any particular individual but is present to some degree in the vast majority of children.

While many gendered behaviors emerge that don't require awareness of gender on the part of the child, there are a lot of gendered behaviors that are purely cultural, and in us humans, these carry a lot of importance (compared to cats or bighorn sheep). This increases the relevance of having a conscious preoccupation with our own gender. The gender identity apparatus in our brain drives this preoccupation.

A good example of this is how we dress. A child's gender identity apparatus will come into play when they are deciding how they want to dress. Another example is role models. Most little girls choose role models who are girls or women based on their own innate sense of being a girl. Most little boys choose role models who are boys or men based on their innate sense of being a boy. They then mimic behaviors of these role models.

There ends up being an interplay between innate preferences and cultural influences. Based on their differentiated brains, girls tend to prefer certain activities. For example, more girls than boys innately prefer social play, so when girls play together, social play will usually prevail. Social-play games for girls will then get more social sanction from parents, teachers, and older peers, which will reinforce this. But, to reinforce it even more, most girls will actively pursue activities that other girls are interested in simply because they are programmed to want to conform to the norms of the gender that they identify with. Their gender identity app will play a role in their conviction of their own gender at this early age and then help drive their interest in "girl" activities such as social play, as well as their interest in female role models.

Don't forget that I am talking about trends. There are plenty of exceptions to these situations, so these descriptions only describe the majority. However, the majority has great power in determining the cultural norms for gendered behavior. Meanwhile, there are many reasons why particular individuals might resist cultural norms, so none of this is monolithic.

Development of gender identity

Now, let's talk about the gender identity apparatus in the brain.

If you have a masculine version of this app, you will have an innate conviction that you are a male. If you have a feminine version of this app, you will have an innate conviction that you are a female.

This app is separate from sexual orientation, and it is also separate from baseball throwing and other gendered apps, but it works in pretty much the same way.

A fetus develops testes or ovaries at weeks 6–8 of gestation based on whether or not an SRY gene is present to activate testosterone production. They subsequently develop genitalia that differentiate to penis vs. vagina/clitoris by week 12 based on whether or not there are high levels of testosterone.

Only later will the various structures of the brain have their critical periods. One of these structures is the baseball-throwing app. Another is the sexual orientation app. Yet another is the gender identity app.

Even before the critical period arrives, the template for the gender identity app is already present in the genes and might be either more or less feminine (or masculine) based on inheritance. Thus, there will be some variability even before the critical period. Then, when the critical period arrives, the male fetuses will mostly have high testosterone and will likely see their app masculinize. The female fetuses will mostly have low levels of testosterone and will likely see their app feminize. Meanwhile, inherited genes code for how sensitive the app will be to the masculinizing effects of testosterone. Some people will have inherited more sensitivity or more resistance to testosterone. Some of those people will happen to have lower or higher testosterone levels on those "critical period" days which will impact the outcome. As always there will be a gradient. And there will also be divergencies.

Most of those who end up with penises and testes will end up with a masculine identity app. Most of those who end up with vaginas and ovaries will end up with a feminine identity app. Some people will be divergent on this trait. Many of those with this divergency will later identify as trans or transgender. Some of those who have versions that are in between masculine and feminine will later identify themselves as non-binary.

In other words, trans people started out at conception with a genetic potential to turn out trans, but it wasn't finalized until later when their brains were undergoing development and differentiation. The differentiation of their gender identity apparatuses happened sometime after week 12, by which time their gonads and genitalia had already formed in the opposite direction. In the case of trans females, there was insufficient testosterone during critical periods to masculinize their gender identity app so it feminized. In the case of trans males, the testosterone present was sufficient to masculinize theirs in spite of their lack of testes.

Gender identity and genes

Gender identity and inherited genes work in a similar way to sexual orientation.

Everybody has every gene needed to have a feminine or masculine outcome for any trait. That means that technically, anybody has a genetic potential to turn out cis or trans. Our gender identity is determined by the interplay of a number of genes, so the particular set of genes somebody inherits will impact whether or not they have a significant potential to turn out trans.

Most people have a minimal potential for a trans outcome. The majority simply have a set of gender identity genes with a more binary response to testosterone. They simply go one way if there is a lot of testosterone and the other if there is not, so they are highly likely to turn out cis. On the other hand, those with a significant potential to turn out trans have a set of genes that lead to a more variable response of their gender identity app to testosterone levels.

For example, if they are an XY fetus, but their developing gender identity app is resistant to the effects of testosterone, then it will only masculinize if there are higher than average levels of testosterone during the critical period for that app. So, some of these XY fetuses will turn out cis, and some will turn out trans. This is one possible mechanism to explain why identical twins might turn out differently. If their critical periods are offset by a few days, and their testosterone levels are fluctuating, then they could end up with different responses and different outcomes.

In the case of XX fetuses, they might end up with a version of that apparatus that is highly sensitive to testosterone. It might be so sensitive that even the levels of testosterone present in an XX fetus might be enough to masculinize that apparatus. Since there is some variability in testosterone levels, some of them might turn out trans if there is enough testosterone during the critical period, while others will turn out cis if testosterone levels are lower on those days.

Unfortunately, there has been much less genetic research around gender identity compared to sexual orientation so far, but hopefully, this science will advance in the coming years.

Sexual orientation, gender identity and correlation with other gendered traits

Since testosterone levels are playing a role in the sexual differentiation of the sexual orientation apparatus, then it seems plausible that it might cause a correlation with other gendered personality traits. The individual male with a feminized version of their sexual orientation apparatus might also have some other divergent traits due to the same process. If their sexual orientation app was feminized because of low testosterone levels on a certain day, then it might be possible that other apparatuses have critical periods at the same time and are also feminized. It is also plausible that whatever caused the lower testosterone action that day might happen several times during the pregnancy. That would also lead to other apps that are more feminized. It is even possible that that person simply has a more globally feminized brain due to a more pervasive lack of testosterone action. That person would have a higher number of feminized apps compared to other males.

The baseball-throwing app is a good way to illustrate this. Based on my observations, most (but not all) straight men can effectively throw a baseball. Most of them have a masculinized app for baseball throwing. Based on my (non-scientific) observation, there are a lot of gay men who can't throw a baseball. But there are many who can. So, in gay men, there are some who have a masculine version of this app and some who have a feminine version.

This has not been well studied, but it does hint that more gay men than straight men have a feminized version of this app. We can infer from this that whatever caused the sexual orientation app to feminize might have also led to the feminization of the baseball-throwing app in this person. For example, they might have had lower testosterone action during the critical periods for both of these apps.

On the other hand, some gay men have a masculinized version of the baseball-throwing app and are very effective ball throwers. During their development, it is possible that the conditions that caused them to have a feminized sexual orientation were no longer present when their ball-throwing app was differentiating. In other words, they might have had lower testosterone levels during the critical period for their sexual orientation app, but during the critical period for the ball-throwing app they could have had higher levels of testosterone, leading to a masculinized version of that trait.

There are also other possible mechanisms. It could simply be the random fallout of genes. Some men might inherit an app for ball throwing that is inherently more masculine, even without testosterone. They might be from a family where everyone inherits a masculine ball-throwing app, including the girls and the gay boys.

However, it would not be surprising if some gay men have a mosaic of traits that is more feminine than the average male. You could say that those men became gay because of a developmental mechanism that also spilled over into other apparatuses. However, we can also recognize that there are gay men who seem to have a highly masculine mosaic and appear to have fewer feminine traits. This would make sense if the developmental pathway that led to their feminized sexual orientation did not spill over and impact other traits.

Let's continue discussing the ball-throwing stereotype, but this time in females. There is a higher percentage of gay women who can effectively throw a baseball compared to straight women (based on my non-scientific observations). It is possible that among the gay women who are effective at throwing a baseball, their baseball-throwing app masculinized due to the same factors that lead to their sexual orientation also masculinizing.

Once again, some of these cases are probably random. Or maybe they just come from one of those families where everyone can throw a baseball. Remember, there are plenty of straight people who don't fit the stereotype either, and there are plenty of straight women who can throw a baseball. From this fact, we can infer that some of the gay women simply got a masculine version of this app due to random inheritance. But considering the trend of gay women being better at throwing a baseball on average, it is likely that some of them got the masculine version because of the action of testosterone during fetal development.

On the other hand, there might be females with a lot of masculine traits, such as ball throwing, but who happen to be straight. Some of these females might have had more testosterone activity during their development, impacting many of their apps, but it just didn't happen to touch their sexual orientation app. Overall, these females might have more divergent apps (masculine apps) due to the actions of testosterone, but it just didn't end up impacting their sexual orientation. This could have arisen from random timing of critical periods but also could be related to inheritance.

In a parallel way, there can be straight men with predominantly feminine traits. It is possible that some of them had decreased testosterone action impacting many of their apps, but it just didn't happen to impact their sexual orientation app. Once again, this could be due to random timing of critical periods, or inherited factors, or both.

In summary: Gay men are more likely to have more feminine traits compared to the average man, but there are plenty of gay men who don't. Gay women are more likely to have more masculine traits compared to the average woman, but there are plenty who don't. This is what you would expect if there are different mechanisms at play for different gay people when it comes to the sexual differentiation of their orientation app.

All of the above discussion also applies to gender identity. In any individual, there may or may not be a correlation with other gendered traits. It makes sense that, on average, there would be more correlation. The mechanism that leads to a divergency of the gender identity trait is probably somewhat more likely to lead to a similar divergency with baseball throwing, or interest in auto mechanics, or even sexual orientation.

However, there is not an exact correlation. A person born with a vagina who has a masculine gender identity may or may not have a masculine sexual orientation. A person born with a penis who has a feminized gender identity app may or may not have a feminized sexual orientation app. However, I would guess that there is a correlation. I would guess that there is a correlation for many other traits as well. This will need more scientific study.

In summary: There are trans men who are oriented towards women and others who are oriented towards men. There are trans women who are oriented towards men and others who are oriented towards women. These traits are likely correlated, but are only partially correlated, which is what you would expect based on the fact that there are probably different mechanisms at work for different trans people.

Remember the gradients

We have to remember all of the gradients. There are people who have in-between gender identities. These people don't identify as solely male or solely female. This is what you would expect from a trait or app that undergoes sexual differentiation. You expect some people to land somewhere between the two extremes. This is true of sexual orientation, of baseball throwing, of interest in ballet, and of gender identity.

Parameters—Fluidity

Something that seems to muddy the waters is fluidity. This refers to people who seem to fluctuate over time with their sexual orientation or gender identity.

Let me explain how that happens.

Each individual has a sexual orientation app, which is a brain apparatus that impacts and controls their orientation. Each individual also has a gender identity app. However, neither of these apps is monolithic. Each individual has their own unique versions with their own unique parameters.

Some of these parameters might not even be impacted by sexual differentiation.

For example, Sexual orientation parameters also include
1) how much sexual aversion that person might have toward their non-preferred gender,
2) their baseline interest in sex,
3) their relative need for novelty vs familiarity with respect to sexual excitement,
4) how much they are attracted to various masculine or feminine traits,
5) etc.

The gender identity app will have a variety of parameters as well. These parameter surely include

1) how much they care about conforming to gender norms in the first place,
2) how much they are influenced by cultural norms in their gender expression,
3) how much influence does their particular gender identity app have on their motivations and behaviors,
4) etc.

An important parameter of each of these apps is the relative fluidity vs. permanence of that trait, in that person. Also, if that trait fluctuates with time, what is the range of that fluctuation.

This can all be confusing because it is shown that people can't really change their sexual orientation or gender identity intentionally. In other words, there does not exist a technique or a therapy that can change your sexual orientation or gender identity. In fact, efforts to do that have been proven to be harmful.

However, there are people whose sexual orientation is fluid. Those people don't really have a change of their basic orientation, but they just have a set of capacities that is more inclusive, and what they are attracted to may fluctuate for reasons that they can't consciously control. People who have fluid orientations have something in common with bisexual people in that they are capable of being attracted to both sexes at some point in their lives. It is possible that like bisexuality, fluidity is also a manifestation of having a sexual orientation app that is somewhere between masculine and feminine.

Fluidity of sexual orientation is also less surprising if we remember that all of us have an evolution of what we are attracted to. For example, most people become progressively more attracted to older people as they themselves grow older. Once again, we don't really control that with intention, and it happens to different extents from one person to another. This typical evolution resembles the fluidity that some people have with their sexual orientation and involves parameters that may or may not relate to gender and may or may not be impacted by sexual differentiation.

It is possible that some people have a similar fluidity with their gender identity. It seems like this would be more probable in people whose version of the gender identity app is not on an extreme end of the masculine/feminine gradient but more in between. In that situation it is more likely that other life factors will lead to some fluctuation during the life of that person. This is consistent with an app that has diverse amounts of sexual differentiation. This gradient emerges as an interaction of genes, hormone levels during critical periods, and other environmental factors.

It is also probable that the gender identity app manifests more strongly in some people and more weakly in others. Some people will identify very rigidly with their gender (in fact, a lot of people). Meanwhile, others might simply have no strong feelings about their gender or might feel non-binary simply because their particular gender identity app has a weaker overall influence.

Chapter 5—Orientation and Identity

However, we also need to be aware that most people, both cis and trans, don't have fluidity of their gender identity but rather have a clear and consistent lifelong pattern.

Parameters—Relative comfort/distress

I mentioned these parameters that impact gender identity:
1) how much they care about conforming to gender norms in the first place
2) how much they are influenced by cultural norms in their gender expression
3) how much influence does their particular gender identity app have on their motivations and behaviors

Since trans people are divergent compared to their assigned gender or compared to their gonad/genitalia configuration, some of them might experience a high level of distress about their bodies or about the role they are feeling forced to play. This can be extreme for many trans people and is referred to as gender dysphoria.

Meanwhile, there are others with divergent gender identities who might self-identify as trans but do not feel as strongly about it. They might be able to navigate life in their assigned gender without substantial distress.

This parameter is particularly important to recognize, because it helps explain the wide variability in how different trans people adapt to this divergency.

Parameters—Some people are asexual, aromantic, or both

Some people have very low attractions to any gender. Others might have high aversions to any gender. These are two parameters that are independent of gender but impact sexual orientation just the same. These parameters can play an important role for people who are asexual or aromantic. However, even people who are asexual or aromantic will likely have a sexual orientation that is on the masculine/feminine gradient because this is a separate parameter—whether or not this is relevant to that person's life.

Review of most important points

1) Different brain structures (and therefore different personality traits) have different levels of sexual differentiation due to genetic makeup and also due to critical periods which are highly impacted by hormones.

2) We are all mosaics. We all have divergencies. We all have both feminine and masculine traits.

3) Divergencies are good for humanity. Variety benefits society.

4) Divergencies arise due to different factors, including inherited genes and environment. A very common contributing factor in divergencies is the fact that sexual differentiation of the gonads and genitals happens much earlier in the pregnancy than sexual differentiation of the brain.

5) Some of us have sexual orientation apps that are divergent compared to our gonads/genitalia. Critical periods of development are a key part of how this can occur, as with any gendered trait. Inheritance gives us the potential for a divergency, and testosterone levels make the final determination during critical periods.

6) Some of us have gender identity apps that are divergent in the same way. Inheritance will give the potential for a divergency, but the testosterone levels during critical periods will make the final determination.

Chapter 6—Chromosomes part 2

Intro

In the next section, "chromosome basics", I will give a very concise overview of how DNA and chromosomes work. This is standard information that many of you learned in biology class. Don't worry if you don't fully understand how chromosomes work. I include this section in an effort to be complete, but for the purposes of this discussion, this is optional information. For visual learners, there are lots of great YouTube videos that explain nicely how chromosomes work.

In the subsequent section, I will describe "exceptions to the chromosomal binary." This discussion is very important and has been in the spotlight recently because of its impact on sporting events. Intersex conditions are rare, but acknowledging they exist and understanding them is also very critical to a comprehensive understanding of gender. Here, I am just giving a brief overview of them, but I encourage everybody to learn more about the diversity that exists within the human chromosomal makeup and how it manifests in human lives.

Chromosome basics

Chromosomes/genes

Chapter 6—Chromosomes part 2

Briefly, human cells have 23 pairs of chromosomes, for a total of 46. These include 22 pairs of autosomes and one pair of sex chromosomes. Each chromosome has genetic code for numerous genes that are all mapped out on specific locations along the length of the chromosome. Humans are estimated to have over 20,000 genes.

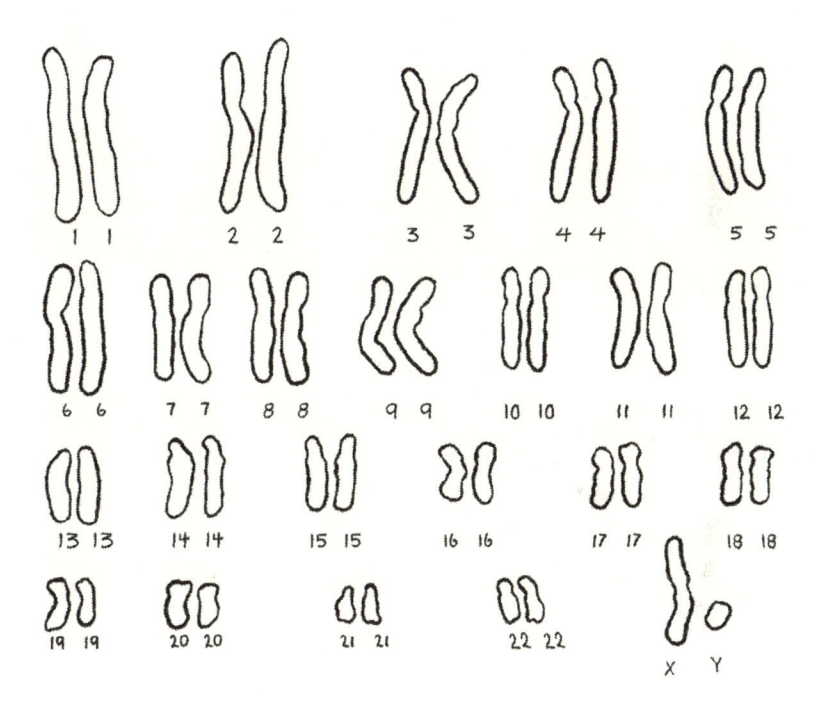

Autosome pairs
There are 22 autosome pairs. Each pair includes two very similar chromosomes that are the same size/length and code for the same genes. However, they also have numerous small differences, which means we have two different versions of almost every gene. Each gene codes for traits that can

133

potentially be expressed in the individual. There are different mechanisms that determine which version/copy of the matching genes will be expressed. It can be one or the other, but usually, both of the gene copies are expressed.

Sex chromosomes—female

Females have two X chromosomes and thus have two versions of every gene on the X chromosome. However, in most of their cells, one of the X chromosomes folds up and is not expressed. However, either of the X chromosomes might fold up in any particular cell. Due to this randomness, all the genes on both X chromosomes end up being available for expression, at least somewhere in the individual.

Sex chromosomes—male

Male humans (and all male mammals), instead of having two similar sex chromosomes, have one X and one Y chromosome. The X and Y chromosomes are completely different. The Y chromosome happens to be much smaller than the X chromosome and codes for a different set of genes. Since males have just one X chromosome, they have only one version of any genes that are found on the X chromosome, compared to females, who have two versions of each of those genes.

The most important gene on the Y chromosome is the SRY gene that activates initial testosterone production and sets a fetus on the path of masculine development. The genes for testosterone and masculine features such as gonads and genitalia are not found on the Y chromosome but are found on the other chromosomes, meaning that females also carry a copy of all of these other "male genes" even if they aren't expressed. Males also carry a copy of all of the "female genes," even if they aren't expressed.

Chromosomes—eggs

Female humans (and all female mammals) ovulate and produce eggs. Each egg will carry one-half of her chromosomes. Since there are 23 pairs of chromosomes, when a female produces an egg, each egg will inherit one of the chromosomes from each chromosome pair. Each egg will carry one of the mother's X chromosomes along with 22 autosomes for a total of 23 single chromosomes. Every resulting offspring, both male and female, will inherit one X chromosome from their mother.

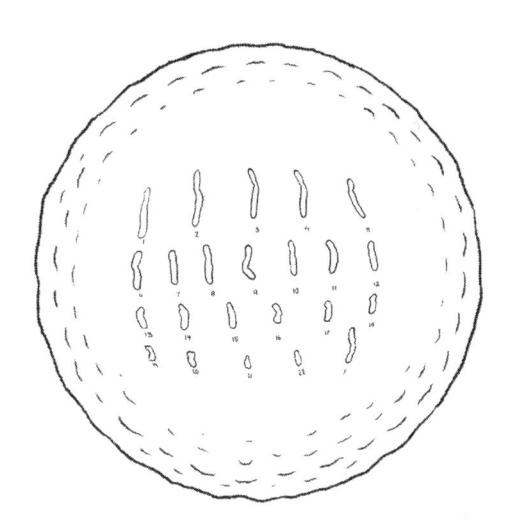

Chromosomes—sperm

When males make sperm, each sperm will have one copy from each of the 22 autosome pairs plus one sex chromosome, either X or Y, for a total of 23 single chromosomes. Since the sperm can carry either an X or a Y, while eggs always carry an X, it is the sperm that will determine the sex/gender of the resulting embryo.

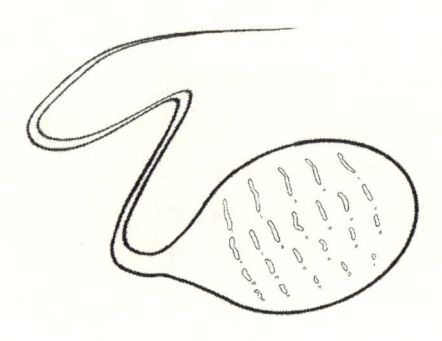

X-linked traits

Since males have only one X chromosome, any trait that is coded for by a gene on the X chromosome is inherited entirely from their mother. Thus, in males, any X-linked trait is also inherited completely from their mother. Daughters get an X chromosome from their father and another from their mother, so in females, each cell has two X chromosomes to choose from. Sons don't get a second X from their father but only get the smaller Y. This means that for all the genes on the much larger X chromosome, males have to rely on getting a healthy copy from their mother, which doesn't always happen—thus the problem of X-linked illnesses in males (such as hemophilia).

Chapter 6—Chromosomes part 2

Exceptions to the chromosomal binary

I am going to briefly describe some exceptions to XX and XY configurations.

XO
There are females who only have one X chromosome and no Y chromosome. Their genotype is referred to as XO instead of XX. This is called Turner Syndrome. These females are more likely to have fertility problems as well as some developmental problems. However, they can survive and live a reasonably normal life, and some of them may never be aware that they have this uncommon genetic makeup.

XXY
There are males who have two X chromosomes in addition to their Y. Their genotype is XXY instead of XY. This is called Klinefelter syndrome. These males also have some very mild developmental issues and fertility issues, but once again, they can have a mostly normal life, and many of these individuals will never realize they have an uncommon genetic makeup.

These two syndromes show us that for some reason females have a more regular development if they have two copies of every gene on the X chromosome, whereas males have a more regular development if they have only one.

XYY
There are males who have an X chromosome but end up with two Y chromosomes. This is called Jacobs syndrome. These

males can also have developmental and fertility issues, but they are less obvious, which makes sense since the extra Y chromosome is small and has very few genes. Most males with they condition will live a relatively normal life and not be aware that they have this uncommon genetic makeup.

Like the X chromosome in males, there is a more regular development if there is only one Y chromosome.

There are other combinations such as XXYY, or XYYY but these are even more rare. There are no cases where the X is missing completely. Humans can't survive without at least one X chromosome.

XX males and XY females

Some very interesting and informative situations also inform our understanding of sex chromosomes.

XX Males
"XX male syndrome" or "de la Chapelle syndrome" is where a male has two X chromosomes and no Y chromosome. Most of the time, this is a result of an SRY gene being present on an X chromosome. Remember that the SRY gene is the most important gene on the Y chromosome and turns on the switch that leads to testosterone production and the development of testes. If, for some reason, this SRY gene is present on one of the X chromosomes, it will do what it does best and turn on testosterone production at the critical time, leading to a more masculine development.

This happens due to a process called crossing over. All of our autosome pairs engage in crossing over. That is where a certain region of a chromosome will trade places with that same region on the other corresponding chromosome within the chromosome pair. This happens frequently in the autosomes, to the extent that the 22 autosomes you inherited from your father get substantially mixed with the 22 autosomes that you inherited from your mother. When you pass on any particular chromosome to your children via egg or sperm, each chromosome will be a mashup of the one you got from your mother and the one you got from your father. That leads to a much higher number of possible genetic combinations.

In females, the two X chromosomes also cross over in a similar way. In males, the regions of the Y chromosome at the very ends can cross over with the very ends of the X chromosome, leading to some minor DNA trading. However, most of the Y chromosome does not cross over, including the part with the important SRY gene. Rarely, however, this part with the SRY manages to accidentally cross over and get implanted into an X chromosome. This X chromosome containing a crossed-over SRY can get passed from a father to an offspring, leading to an XX individual who has the SRY gene on one of their X chromosomes.

XX males can end up with a configuration that is somewhat similar to Klinefelter's (XXY). They have two X chromosomes, but they also have the most important part of the Y chromosome.

In these cases where there is an SRY gene on the X chromosome, this SRY gene can be fully functional, or non-functional, or somewhere in between. If it is fully functional, that person will have their testosterone factory turned on and they will develop a penis and testicles, and their female genes will be suppressed. If the SRY gene is non-functional, then they will develop in a regular female direction. If the gene is partially functional, then they will proceed toward an in-between type of development where they might have both female and male genitalia, both of which will likely be partially developed.

So, you don't really need a Y chromosome to become a male (although it helps). It seems that all you really need is an SRY gene. But even more fundamentally, what is really needed to induce masculinization is enough testosterone delivered at the right time. This is what the SRY gene instigates.

Another situation where XX individuals masculinize occurs with "congenital adrenal hyperplasia." In this syndrome, the adrenal gland produces excess androgens (similar to testosterone) that induce varying degrees of masculinization in the fetus. This can lead to ambiguous genitalia due to clitoral enlargement. This is the most common cause of ambiguous genitalia among XX individuals.

XY Females
If a fetus has an XY genotype, but their SRY gene is ineffective or inactive due to a mutation in that part of the chromosome, then that fetus would develop as a female similar to an XO female.

Another interesting condition is called "Androgen Insensitivity Syndrome." Some people with XY chromosomes have a resistance to the effects of androgens, including testosterone. This can happen in various degrees, but those with the most extreme insensitivity will not develop male characteristics in spite of having an XY genotype and in spite of the presence of large amounts of testosterone. These individuals will appear female, and many will never find out that they have this syndrome, although they will not be fertile. Those who have a less extreme insensitivity to androgens might end up with only partial male development.

Intersex

An intersex person will have chromosomes, and/or gonads, and/or genitalia that cannot be clearly categorized as male or female. Some of the conditions I described above would fit under the umbrella of intersex conditions. These conditions have been observed throughout history but only recently are we starting to understand the developmental pathways that lead to intersex outcomes.

Chapter 7—Epilogue

Overall summary

The six most common systems for classifying gender are
1) *whether that person appears to have the characteristics of a person who is/was/will be able to become pregnant vs. a person who is/was/will be able to impregnate somebody,*
2) *whether or not that person has a penis vs. a vagina,*
3) *whether that person outwardly appears to be male vs. female based on presentation and visible traits,*
4) *whether that person contributes a larger gamete (e.g., egg) vs. a smaller gamete (e.g., sperm),*
5) *whether that person has XX sex chromosomes vs. XY chromosomes,*
6) *whether that person has an internal sense/awareness/conviction of being female vs. an internal sense/awareness/conviction of being male.*

Definitions for how certain terms are used in this book:
A feminine trait is any trait that is more common among females than males.
A masculine trait is any trait that is more common among males than females.
Divergent traits or divergencies are any masculine trait in a female or any feminine trait in a male.

Chapter 7—Epilogue

We all carry in our genes all the information necessary to make both a complete female and a complete male with the exception of the SRY gene which is found on the very small Y chromosome that (generally) only males carry.

Sexual differentiation starts with our genotype, which is determined at conception. Males (generally) have an XY genotype and females (generally) have an XX genotype. The Y chromosome is very small and only has a few genes. The most important Y chromosome gene is the SRY gene that turns on (or activates) testosterone production. When this happens, the gonads become testicles and start secreting high levels of testosterone, which continues throughout the rest of the gestation. If there is no SRY gene, then the testosterone machine is not activated, and the gonads become ovaries. After gonad development is established by week 8 of gestation, the genitals will masculinize or feminize depending on the levels of testosterone. The penis or vagina is thus formed by week 12. Other reproductive structures such as the uterus and prostate also form during this period.

During the rest of pregnancy, the brain continues to develop layer by layer. As each layer develops, it passes through critical stages of development where it is subject to signals that will determine its pathway.

Every brain apparatus (that is subject to the effects of testosterone) will go down a path of masculinization or feminization based on the testosterone levels that are present during a critical period, generally early in the development of that brain tissue. This will determine where that trait will end up on the masculine/feminine continuum.

These apparatuses/traits in the brain have critical periods that happen at different times and happen later in gestation compared to the development of gonads and genitalia.

Brain development continues well into adulthood. It continues to have further sexual differentiation during some of these later stages, especially at puberty.

Since there are fluctuating levels of testosterone, then each tissue might have a different level of sexual differentiation depending on the levels of testosterone that happen to be present during the critical period for that tissue. Other factors that might be present can also impact the outcome, such as toxins, maternal hormones or antibodies.

The template for each trait is coded in the DNA and is inherited. Any template can be intrinsically more feminine or masculine even prior to the effects of testosterone levels. The DNA will also dictate how much each tissue will respond to the presence (or absence) of testosterone. Each trait is thus a result of nature AND nurture—in this case, the DNA is the nature, and the testosterone levels are the most important part of the nurture.

Due to (1) critical periods and (2) inherited factors, as described above, many of these traits or apparatuses will be

Chapter 7—Epilogue

divergent. When I say divergent, I mean a feminine trait in a male or a masculine trait in a female. These divergencies are a normal part of human variation, and every single person has numerous divergent traits.

We have numerous traits that are based in the brain. This includes all of our personality traits. It also includes all of our traits that determine our interests, attractions, skills, emotional responses, etc. Each trait is the result of an apparatus in the brain that dictates that trait. Each apparatus is spread out in diffuse regions of the brain and includes its unique interconnections.

Each trait is like a smartphone app and each "human app" ("apparatus") works by coordinating the brain activity of the relevant brain cells in these diffuse regions. These apps take in relevant sensory information from our sensory organs, process it, and then send out signals to our muscles, organs or other parts of our brain.

We have apps for all of our interests, attractions, capabilities, emotions, etc. Some of these apps were described earlier, including ball throwing, interest in ballet dance, interest in auto mechanics, empathy, language, etc. A large number of these traits/apps are gendered, and the process for how they undergo sexual differentiation is the following:

1) We have a tissue in the body or brain that has the potential to eventually develop into a structure/apparatus that is either masculine, feminine or somewhere in between.

2) At a certain stage in development, with precise timing, that tissue will begin development and differentiation.

3) At a critical period during development (usually early), that tissue is highly sensitive to the presence or absence of high testosterone levels (working in balance with other hormones). The presence of higher testosterone levels during this period will set that tissue (or trait) on a path toward masculine development. The absence of high testosterone levels will set that tissue (or trait) on a path toward feminine development. There can also be intermediate outcomes. In fact, when it comes to brain traits, most of the outcomes are intermediate.

4) Other environmental factors will also play an outsized role during critical periods (e.g., toxins, antibodies, etc.).

5) Inherited genes are the blueprint. Genes guide and individualize each step in the process. Genes give the initial template that can be intrinsically more feminine or masculine even prior to differentiation. Genes also determine how each trait/tissue will respond to testosterone during critical periods. Genes can also impact the final outcome of any trait through a variety of mechanisms, some of which are independent of testosterone.

Our gender identity app determines our internal sense/conviction of our own gender. This app plays a large role in our lives and behaviors starting at a young age. Other animals, especially mammals and birds, have their own versions of gender identity apps that drive their gender-specific behaviors and groupings. However, the gender identity apps in humans require much more conscious awareness. This is important because all human societies have gender roles, but many of the roles are completely cultural and specific to that particular society. At a young age, most children develop a sense of their gender identity and also a sense of the "correct" behaviors for their gender in that specific society, and they are drawn toward those behaviors/activities/interests. While many of these gender roles are completely cultural, others are an outcome of hard-wired brain traits that have an average gender difference. This average difference ends up being amplified by the expectations of society and the general desire among most individuals to conform to gender norms.

Gender identity apps can be divergent. A person born with a penis could have a feminized gender app and identify as female. A person born without a penis could have a masculinized gender app and identify as male. These people might be referred to as trans.

Sexual orientation can also be divergent, and this is common in humans as well as some animals. A male with a feminized sexual orientation app will be sexually oriented toward males and will be referred to as a gay male. A female with a masculinized sexual orientation app will be sexually orientated toward females and will be referred to as a gay female or lesbian.

Every gendered trait will differentiate to some extent but most of them will only differentiate to versions that are neither extremely masculine nor extremely feminine. Most traits will be on a gradient between masculine and feminine, especially traits that are based in the brain.

Sexual orientation and gender identity can also fall on the continuum. People who fall in the middle of the gradient of sexual orientation could be considered bisexual or pansexual. People who fall in the middle of the gradient for gender identity could be considered non-binary.

The factors that lead to divergency of sexual orientation or gender identity may (or may not) spill over and lead to divergency of other traits (correlation). This is what would be expected based on how these traits develop.

This is a summary of how sexual orientation and gender identity develop, as well as all other traits that undergo sexual differentiation:

1) *Inherited DNA codes for the genes and gives a basic undifferentiated template for that app, but even before being acted on, it may be intrinsically more feminine or masculine based on inheritance.*
2) *The regions of the brain that are involved with this app will each pass through critical periods. Critical periods of sexual differentiation in the brain happen after the gonads and genitals have already formed.*
3) *During critical periods, the relevant tissues are particularly sensitive to testosterone levels, which will lead that tissue to masculinize or feminize, depending on the testosterone levels.*

4) *Other environmental factors (such as other hormones, nutrition, toxins, and antibodies) will also have an impact, especially during critical periods.*
5) *After birth, the individual's external environment will continue to act on this app, including the physical environment, and also culture, learning, parenting, etc.*
6) *After birth the individual's internal environment (hormones, etc.) will also continue to play a role.*

All of this will lead to each person having a unique mosaic of masculine and feminine traits. This includes everybody—straight or gay, cis or trans. Furthermore, most of our traits will fall along the continuum of partly feminine and partly masculine, which will further increase the variety among humans. However, due to the action of testosterone and the subsequent role of culture and learning, most males will have mostly masculine traits and most females will have mostly feminine traits. But there will be lots of exceptions and no human brain will be completely feminine or completely masculine.

Closing thoughts

1) Some of these details will eventually be proven, and some will likely be disproved. So far, there has been a lot of research, and we now understand better than we ever have, but I look forward to more enlightenment from future research. I also hope that we can pinpoint any inaccuracy I might have made and get even more specific details about these developmental pathways.

2) I hope I have made effective use of stereotypes in my examples. I also hope that I made it clear why nobody's destiny is set by these stereotypes or trends. Everyone of us has some stereotypical traits, but all of us have traits that defy stereotypes.

3) I hope I was able to simplify some of the complicated concepts to promote a reasonably accurate understanding for a broad audience. There will be experts out there with a more nuanced understanding of these highly complicated processes, and I look forward to their feedback.

4) I plan to update this book every few years so that it can include any new developments in our knowledge as well as any new information that comes to my awareness. This version was written in 2024, so watch for a newer edition if some years have passed since then.

5) I am hoping I can complete an additional reference that will cite the most relevant research and expert opinions on this subject, including articles, studies, video links etc. and include that as a section in future editions.

6) Last of all, I am hoping that I can also complete a version of this book with more illustrations to benefit people who are more visual learners (or maybe even a video series).

About The Author

Daniel Y. Parkinson, MD, is a physician and a psychiatrist. In writing this book, he drew on his expertise acquired from his medical training, psychiatric practice, and extensive review of the research. His goal is not to pass judgment on these issues or prescribe policy but rather to increase understanding, which will lead to better outcomes for our lives, our families, and our societies.

Made in United States
Troutdale, OR
12/14/2024